The Future of Television - Convergence of Content and Technology

Edited by Ioannis Deliyannis

Published in London, United Kingdom

IntechOpen

Supporting open minds since 2005

The Future of Television - Convergence of Content and Technology
http://dx.doi.org/10.5772/intechopen.75322
Edited by Ioannis Deliyannis

Contributors
Henrik Vejlgaard, Yasser Ismail, Laura Camila Ramirez Bonilla, Ananda Mitra, Ioannis Deliyannis

Notice
Statements and opinions expressed in the chapters are these of the individual contributors and not
necessarily those of the editors or publisher. No responsibility is accepted for the accuracy of
information contained in the published chapters. The publisher assumes no responsibility for any
damage or injury to persons or property arising out of the use of any materials, instructions, methods
or ideas contained in the book.

First published in London, United Kingdom, 2019 by IntechOpen
IntechOpen is the global imprint of INTECHOPEN LIMITED, registered in England and Wales,
registration number: 11086078, The Shard, 25th floor, 32 London Bridge Street
London, SE19SG – United Kingdom
Printed in Croatia

British Library Cataloguing-in-Publication Data
A catalogue record for this book is available from the British Library

Additional hard copies can be obtained from orders@intechopen.com

The Future of Television - Convergence of Content and Technology
Edited by Ioannis Deliyannis
p. cm.
Print ISBN 978-1-78985-753-5
Online ISBN 978-1-78985-754-2

We are IntechOpen,
the world's leading publisher of
Open Access books
Built by scientists, for scientists

4,000+
Open access books available

116,000+
International authors and editors

120M+
Downloads

Our authors are among the

151
Countries delivered to

Top 1%
most cited scientists

12.2%
Contributors from top 500 universities

Interested in publishing with us?
Contact book.department@intechopen.com

Numbers displayed above are based on latest data collected.
For more information visit www.intechopen.com

Meet the editor

Dr. Ioannis Deliyannis is an assistant professor at Ionian University in Corfu. He is a member of the Faculty of the Department of Audio and Visual Arts and a founding member of the InArts Research Laboratory. He has created various interactive multimedia systems ranging from experimental television stations featuring multiple modes of delivery to educational and multisensory games. He is the author of a series of journal and conference publications in the above field and a series of books targeting the experimental and creative aspects of the technologies involved. He is involved in the design of user-centered software products and services, focusing on the use of mobile sensory systems to create intelligent interactive systems, entertainment education systems, educational applications for people with disabilities, multimedia adapters, holograms, interactive navigation narrative applications, and augmented and virtual reality systems.

Contents

Preface

Today, television convergence is a particularly difficult process to address and identify clearly because technologies, content, and users evolve simultaneously, consuming content from all technologies and platforms at the same time. Hence, to the inexperienced observer the future of television seems unclear because we use both old and new media in our everyday lives. This book addresses exactly this issue and attempts an interdisciplinary approach to examine the problem in hand. By reading through the chapters it becomes apparent that the Internet as a broadcasting and interaction medium provides far more capabilities than television, which despite its smart functionality does not provide the user with the ability to process, reproduce, and publish content that can trigger further interaction. As a medium, television remains static with little or no functionality over content control. This clearly indicates that the transformation of television is imminent and it is also a recurring phenomenon. The book focuses both on the technological side of developments, particularly on the advances that enable highly efficient broadcasting to be implemented, while it also examines the social characteristics of the convergence process. This interactive evolution cycle that can be explored only be taking into consideration the dynamics imposed both from technological and social developments.

I would like to thank all the colleagues from the Interactive Arts Research Laboratory at Ionian University for their support in the reviewing process. https://inarts.eu/en/lab/staff/

Dr. Ioannis Deliyannis
Assistant Professor,
Department of Audio and Visual Arts,
Ionian University,
Corfu, Greece

Section 1

Historical and Technological Aspects of Convergence

Introductory Chapter: Convergence of Content and Technology - The Role of Interaction

Ioannis Deliyannis

1. Introduction

Convergence is a transient and multifaceted process that can be examined from various perspectives and timescales [1, 2]. Studying convergence allows us to appreciate changes that occur in our proximity and how our reality is influenced by changes that occur globally [3]. Examining the phenomenon of convergence is a process that usually requires observance of global changes in order to interpret local changes.

The process of convergence is based on a transient mechanism that creates chain reactions affecting a wide variety of domains [4]. If we examine television as a closed system, we can see how those extend from technological to cultural, educational, and social. Take for example the case where advances in technology were significant and able to trigger change: the adoption of HD, 4K, and 8K resolution by the technological manufacturers. This change in resolution significantly affects the transmission process as more information needs to be transferred, the hardware has to be upgraded in order to process more information, and the software has also to be altered in order to support the new resolutions and enable viewers to access the content. The content production perspective has to adapt as well in order to utilize the higher-resolution imaging of the medium, a process that includes an upgrade across the complete content production workflow. The availability of new ultra-sharp content affects consumers who upgrade their viewing and content storing equipment. Other sections that have to adapt include the gaming industry, which needs to offer gaming experiences that support the new resolution, a chain reaction that pushes upward the processing requirements of video cards used in gaming consoles that traditionally use the television as a gaming monitor. Also, Internet-based streaming of content such as IPTV needs to be updated to support the new resolution, while the networks have to be able to sustain the increased data rate [5]. Following this discussion, one may now appreciate that changes that occur at other domains with no apparent direct connection to the domain of television such as a phenomenally simple change of broadcasted image resolution may affect the domain itself.

However, this evolutionary process is not new and is certainly not limited to television, as in the past, the simultaneous introduction of new media technologies enabled a lower-quality standard to evolve instead of the clearly superior one, for marketing reasons [1]. After the introduction of the VHS and BETACAM systems in the market, the lower-end VHS systems won the competition and infiltrated the market as consumers chose to adopt VHS technology due to the wider content variety available [6]. As a result, only the television studios adopted BETACAM systems in order to support their content mastering tasks.

Clearly, convergence cannot occur without interaction between the key players. It is important therefore to define what interaction is and identify its role within the convergence process. The commonly accepted definition of interaction is described as "*a mutual influence or action between two or more agents.*" Here, for the purpose of our investigation, we consider a triplet that involves content, technology, and consumers. They all have an active role in this process and this organization falls within the definition of interaction. We start our investigation with the introduction of the new medium and then follow up its evolution in relation with the other two parts: content and consumers. The same method can be applied to the new medium that seems to be used as the main platform to broadcast and access multiple types of content, including television-oriented: the Internet. Examining how users and content adapt and evolve through the new medium reveals the transformation processes that occur and drive the convergence process.

2. Television and Internet evolution

Television broadcasting networks were introduced at different times worldwide with US, UK, France, Canada, Australia, Germany, Poland, the Soviet Union, and Japan to support adoption until the 1940s [6]. Similarly, this was also the case with the introduction of the Internet communication network until its worldwide adoption as a platform for the support of communication that was concluded in 1993 with the introduction of NCSA Mosaic WWW browser, as it was ported to multiple platforms and was able to display multimedia content.

However, there exist great differences between the two content access platforms. We focus at their broadcasting architecture differences. Television supports *one-to-many* content transmission while the Internet support a *many-to-many* communication architecture offering advanced capabilities as it is classified as a higher-order architecture compared to that of television broadcasting. This is easily explained as the Internet can emulate the transmission method employed within the television architecture, while the opposite is not possible. As a result, television evolved as a closed one-to-many broadcasting system where only few had control over the content that was broadcasted, while the openness of standards and the connectivity capabilities offered by the Internet communication platform allows anyone to broadcast their own content.

However, here the role of the user is highly important and can function as a catalyst for the medium. It took many years after the introduction of the Internet for users to catch up and share their own content at a massive scale [7]. Note here that a basic condition in order to create what is called the "*information society*" is that all the users should possess the capability to contribute content. This clearly was not the case until the introduction of the World Wide Web (WWW), suitable content encoding standards that support content broadcasting and delivery [8]. Various factors may be identified for this negative development including the technical complexities of the task, the nonuniform encoding and broadcasting standards, the slow network speeds offered for uploads through the networks offered by service providers to home users, etc. However, those problems are solved today and in addition, the introduction of the social networks has sparked the public's interest into new uses of the communication networks which is more personalized mobile and allow them to share their daily activities, thoughts, actions, and ideas with particular groups of interest. Another sign of this disorganization is the fact that up to this moment, the traditional communication models do not cover the case of social networks and their interaction complexities. As such, they are only extended to interactive TV modes which still use the one-to-many broadcasting model, supporting limited user interaction.

3. Conclusion: the future of television

Three key players, technology, content, and users affect the future of television. On the technological side, Quality of Service algorithms allow for the delivery of content at varying data rate and resolution, optimizing and automating the content delivery process. Torrent-based streaming solutions allow networks to provide fast live and on-demand content streaming without overloading the networks. It is the evolution of the Internet has reached a point where it features advanced interactivity capabilities and the tools that can support fully interactive information exchange and reproduction. As such, it provides a solid developmental platform for convergence.

From the developer's perspective, the above technological features permit highly interactive narrative scenarios to be implemented and distributed commercially. Yet the most important characteristic here is that normal users become content developers, changing their user-experience as they ultimately gain control over the content that is broadcasted. Social media such as YouTube enable them to create their own thematic channel. They are in a position to even create a media broadcasting station with the use of YouTube Live that provides live broadcasting tools and content archival services. At the same time, cloud-based editors may be used to directly edit and rebroadcast the revised content, while various tools are already designed to automate the capturing and editing process. Support for new content types such as interactive 360° videos are also embedded within social media platforms, enabling users to record, edit, and create their own interactive multimedia experiences. In conclusion, everyday users have today the ability to become content providers, a development that is highly important for the convergence process. However, there is one significant step that is required to complete the convergence process and shape the future of television, as the unification of content providers with the users has created a wide plethora of content sources and the real-life problem faced today is to select which content to access. This requires the use of artificial intelligence in order to provide a unique, entertaining [9], educating [10], gamified [11], useful [12], fulfilling, and easy [13] interactive content-accessing experience to each user.

Author details

Ioannis Deliyannis
Interactive Arts (inArts) Research Laboratory, Department of Audiovisual Arts,
Ionian University, Corfu, Greece

*Address all correspondence to: yiannis@ionio.gr

IntechOpen

References

[1] Deliyannis I. Converging multimedia content presentation requirements for interactive television. In: Media Convergence Handbook, Media Business and Innovation Series, Journalism, Broadcasting, and Social Media Aspects of Convergence. Springer-Verlag Handbook; 2015

[2] Jenkins H. Convergence Culture: Where Old and New Media Collide. New York University Press; 2006

[3] Campbell R, Martin C, Fabos B. Media Essentials: A Brief Introduction. Bedford/St. Martin's; 2018

[4] Syvertsen T. Challenges to public television in the era of convergence and commercialization. Television & New Media. 2003;**4**(2):155-175

[5] Biczók G et al. Towards multi-operator IPTV services over 5G networks. In: IPTV Delivery Networks: Next Generation Architectures for Live and Video-on-Demand Services. 2018. pp. 283-314

[6] Abramson A. The History of Television, 1880 to 1941. Jefferson, NC: McFarland; 1987

[7] Deliyannis I. Information Society and the Role of Interactive Multimedia (in Greek). 2nd ed. Athens: Fagotto Books; 2010. p. 214. ISBN: 978-960-7075-99-4

[8] Deliyannis I. Adapting interactive TV to meet multimedia content presentation requirements. International Journal of Multimedia Technology. 2013;**3**(3):83-89

[9] Deliyannis I, Karydis I, Karydi D. iMediaTV: Open and interactive access for live performances and installation art. In: 4th International Conference on Information Law (ICIL2011). 2011

[10] Kaimara P, Deliyannis I. Why should I play this game? The role of motivation in smart pedagogy. In: Didactics of Smart Pedagogy. Springer; 2019. pp. 113-137

[11] Deliyannis I, Lygkiaris M. Game Development (in Greek). Fagotto Books; 2017. ISBN: 978-960-6685-75-0

[12] Papadopoulou A et al. Art didactics and creative technologies: Digital culture and new forms of students' activation. In: Digital Culture & AudioVisual Challenges Interdisciplinary Creativity in Arts and Technology, DCAC International Conference. Corfu, Greece: Ionian University; 2018

[13] Deliyannis I, Karydis I. Producing and broadcasting non-linear art-based content through open source interactive internet-tv. In: Proceedings of the 9th International Interactive Conference on Interactive Television. ACM; 2011

Television Reception and Technological Convergence in the 1950s: The Case of Mexico City

Laura Camila Ramírez Bonilla

Abstract

The aim of this chapter is to characterize and analyze the television audience in Mexico in the 1950s through the confluence of three audiovisual communications technologies: cinema, radio, and television. This research understands the convergence as a context and a social and cultural experience that changed roles, routines, perceptions, and stereotypes of the contemporary societies. How did the TV audiences of the Mexico City react to the technological convergence? How did cinema, radio, and television become interrelated? What elements mediated in television reception in the 1950s? To answer these questions, this research used the national press and specialized magazines of the time, oral history, and photographic archives as primary information sources. The chapter argues that the audience established a bond with TV from its past experience with cinema and radio. With the reference of the moving image, the immediately transmission and the domesticity, the viewers recognize themselves and react to screen contents. In 1950s, the audiences are active and interactive. This research concludes that technological convergence, from a historical point of view, constitutes a context of transformations and a set of assimilation, integration, and interaction practices. Likewise, it considers that in the middle of the XX century, the world experiments a "early convergence" of contents and technology, where the television is main characters.

Keywords: television, audiences, technological convergence, technological transformation, Mexico City, reception, history of mass media

1. Introduction

Television reception is a subject that has been relatively unexplored from a historical perspective. The same has happened with technological convergence. The past allows us to understand the key drivers of the emergence, change, and current character of television and its audiences, its association with other technologies, and its relationship with historical contexts. In Latin America, this subject was overlooked for various reasons: the difficulty to quantitatively measure the preferences and consumption habits of TV viewers before the 1970s; the reduced spaces where spectators could publicly express their views on media products; the difficult access to archives of TV companies—private and public—and their oldest audiovisual material; and finally, the absorbing trajectory of television, as a media that captured the attention of scholars and neglected the other players in the broadcasting process.

The aim of this article is to characterize and analyze the television audience in Mexico in the 1950s through the confluence of three audiovisual communications technologies: cinema, radio, and television. This research understands the meeting of these media as a context where viewers, still "neophyte" and expectant, recognize themselves and react to screen contents. In Mexico, the technological convergence was key in the formation of the first generation of TV viewers. The audience established a bond with TV according to its past experience with cinema and radio. It is therefore not surprising to understand this convergence as a social and cultural experience that changed roles, timeframes, tastes, perceptions, and social habits, while reinforcing the interactivity of audiences, as well as their relationship with the media and their response capabilities.

Television changed the way to "see" and "listen to" information and entertainment. Three elements were key to shape the way in which the audience understood the arrival of TV within the framework of technological convergence: 1) the city was defined as the place of inception and development of television; 2) both the TV set and television as an industry were directly linked to the idea of progress and modernization in Mexico; and 3) the mass-consumption of technological innovation became a must for commercial advertising, political propaganda and advocacy, communication of education and culture, and even social prestige. These three elements are necessarily interrelated.

This research used as primary information sources the national press and specialized magazines of the time, oral history, and photographic archives (Casasola and Hermano Mayo). The historical narrative made it possible to identify behaviors of both television and recurrent viewers in the decades that followed, in terms of simultaneity, the establishment of routines and interaction, and regarding practices that will rarely reoccur in later years: the community sense, amazement, and recursiveness. How did the Mexico City audience react to the technological transformations and confluences? How did cinema, radio, and television become interrelated as the broadcasting means of the time? What elements mediated in television reception? To answer these questions, this chapter is divided into four sections: the first deals with the concept of television reception, from theoretical and historical perspectives; the second characterizes the TV audience of Mexico in the 1950s; the third focuses on technological convergence as a fact that left an indelible mark on the experience of the TV spectators of the time; and the fourth sets out conclusions on the ownership of a historical study to comprehend the future of television and technological convergence.

2. On TV reception

As reported by Orozco, reception is an interaction mediated by multiple sources, characterized by material, cognitive, and emotional contexts [1]. Audiences provide a unique meaning to the messages received according to their own cultural backgrounds and contexts [2]. They are far more than passive players or mere receptacles of everything displayed on screen [1]. Quite the opposite, as they react: nod, adapt, resist, condemn, or replicate. The practice to portray the viewer as a passive user has dominated both theory and historiography. As pointed out by Castells, the issue with such propensity lies in that the viewer continued being perceived as the object rather than as "the subject of communication" [3].

For Huertas, history has considered and studied audiences from three perspectives: as a mass—a common vision between the eighteenth century and the first half of the twentieth century—focused on the urgency to a fast and immediate communication to the crowds; as diverse groups, where the conformation of markets and differentiated consumers is privileged, as in the case of television; and finally, an individual-focused approach, which since the 1980s has made individualized or customized production a priority—not necessarily translating into better quality of contents—[3].

The 1950s and early 1960s witnessed the emergence and evolution of the viewer in Mexico. The spectator of the 1950s was an active agent, who by virtue of his/her cultural, religious, political, and social background, provided a variety of meanings to TV messages. This phenomenon comprises a whole communication process. Reception should not be understood as a mechanism that takes place in isolation from the media or the message. The agents of this system, even in the precariousness of the new technology, met and shared ideas ranging from curiosity to expectations. It was a tough issue. With TV, the act of communicating to the mass public became complex for three reasons: (1) the moving image, synchronized with sound, could be transmitted live, immediately, from the place of events; (2) the new media had a domestic character, as it no longer required attendance to a public hall to receive the message; and (3) the sense of rituality produced by TV, either at home or in a public space, was novel and favored congregations.

Nonetheless, the television viewer should not be idealized. "This is not to say that the subject of communication is not influenced, or even deceived, by the content and format of the message," says Castells. The construction of meaning is complex and depends on "triggering mechanisms" that combine various levels of involvement in the reception of the message ([2], p. 179). All the links of the communicative process were connected when the screens were turned on. From then onward, each viewer can be understood as an interlocutor with other messages, audiences, and issuers. As a result, viewers are not uniform agents. The audiences of the 1950s, diverse as they were, mirrored the euphoria, curiosity, and admiration of novelty, but also skepticism, rejection, and disenchantment given some of the outcomes and the activism of others in demanding changes. While some received the contents from immobility, with no further questioning, others scrutinized them and expressed sharp criticism [4].

3. The TV viewer of the 1950s

In the midst of an industrializing and modernizing boost, TV officially was inaugurated in Mexico on September 1, 1950. This day, the IV Government Report of President Miguel Alemán Valdés (1946-1952) was read. The government opted for a private and commercial model, similar to the US. The first television concession was granted to the entrepreneur Rómulo O'Farril, to manage Channel 4 (XHTV). The two following years, another two channels were inaugurated: XEW-TV Channel 2, owned by Emilio Azcarraga Vidaurreta, and XHGC Channel 5, by Guillermo Gonzalez Camarena [5]. In 1955, Telesistema Mexicano S.A. acquired control of the three television networks, thus establishing a monopolistic scheme, in the hands of Azcarraga, who established the consortium Televisa S.A. in January 1973 [6]. Only until 1959 the first public project emerged, with an educational and cultural project named Channel 11.

In the 1950s, the TV set was a sumptuous appliance. "When TV sets become affordable for me, I may buy one," said Gabino Granados, a photographer of Alameda Central, when interviewed by the magazine *TV-56,* almost 6 years after the media was inaugurated [7]. Data on the presence and expansion of televisions in Mexico City are scarce and sometimes contradictory. A few days before its opening, O'Farrill, owner of the first channel, predicted that by mid-September 1950 there would be 2000 TV sets across the country, and 10,000 would be spread in households by the end of that year [8]. In 1954, the UNESCO report on information technologies, carried out for 52 state entities, estimated that some 50,000 TV sets were available in Mexico, that is, one for every 578 inhabitants. The statistics estimated a potential of 9 million viewers in the country, equivalent to 34% of its population. When figures were contrasted, it was found that the United States was the leading television power worldwide, with one TV set for every 7.5 inhabitants and 65% of the population considered as TV viewers [9].

By 1958, Mexico had 11 television stations in operation, distributed across several states of the country. An astonishing growth followed in 1959, the country had 16 stations [10]. The UNESCO report for 1963, which detailed the evolution of the TV phenomenon between 1950 and 1960, revealed that in 1958 there were about 184,000 TV sets in the country, equivalent to 6 sets by every 1000 inhabitants. In just 2 years, in 1960, the number had grown to 650,000, equivalent to 19 sets per thousand inhabitants, three times the number recorded in 1958. By then, Cuba was the country best positioned in Latin America, above Brazil and Argentina [11].

3.1 Public, semi-public, and private profile of the viewer

What were the most recurring daily practices of TV viewers in the 1950s? How was TV watched? Who congregated in front of screens? Watching television in the 1950s in Mexico was a public-private entertainment and socialization activity. TV was not a top-level source of information. The behavior of spectators, although diverse, was dominated by curiosity, the aspiration to own a TV set, and expectations regarding further advances to come. The following sections address viewers of the 1950s in public, private, and semi-public settings that allow to delineate their profile.

3.1.1 The public space

Commercial displays were one of the first contacts of Mexicans with *the TV set*, as a technological appliance, and with *television*, as an audiovisual communication system. Store displays were used as an information and advertising strategy. The public knew of the new technology and started to contemplate the idea of acquiring it. Department stores and manufacturers of TV sets organized exhibitions even before the system was introduced. Unintentionally, these exhibits were one of the first "public television watching practices." In Mexico, the photographs by the Mayo brothers, in the *Archivo General de la Nación* (Mexico's General Document Archives), recorded a group of people gathered in a projection hall in front of a *Magestic* TV set displayed in a small shelf [12]. Afterward, the TV set passed from exhibition halls to display cases. On August 13, 1950, *El Palacio de Hierro*, a famous department store, invited its customers to a "practical demonstration" in the basement of its annex building. "The day has come!" it announced in an advertisement published in the *Novedades* newspaper. The message was clear as to the target public and the type of experience it intended to advertise: "Your home can now be the stage of any show... as you and your family can enjoy the result of years and years of research" [13] (**Figure 1**).

Advertising campaigns were adapted to streets, becoming familiar for urban passers-by. "Watching TV" through a showcase from the street was also a way to become a viewer. "I had watched TV in the stores that sold them. There was always an exhibit for people to appreciate TV sets. I remember the *General de Gas* store, located at *Insurgentes*, where a TV set was always on display. That was my first contact," says a lady interviewed in Mexico, who owned a TV set since 1951 [14]. Only a few days before the television system was launched, the *El Nacional* newspaper raffled an *Admiral* TV set among its subscribers. The appliance was displayed in the windows of its corporate offices at the *Alameda Central*, in Mexico City downtown. Passers-by, both adults and children, gathered in the street to admire the new technology [15] (**Figure 2**).

The urban space was filled by curiosity, the experience of watching the small screen working, and the desire to purchase the appliance. The spectator was discovering himself at the very basics: "watching TV," an elementary modality mediated by the technological expectation or by the mere chance of coming across "the wonder" of the time. The practice was arising as a public experience, as it was taking place in a collective, open, freely available space, as pointed out by McQuail ([16],

Figure 1.
Publicity of El Palacio del Hierro. Novedades, Mexico, August 13, 1950.

Figure 2.
El Nacional showcase in Mexico City. In: El Nacional, September 13, 1950, Mexico.

p. 26), but at the same time was tied to the control by a private stakeholder—a third party— who owned the appliance and decided when to turn it on and expose it to the public. In these dynamics, the casual passer-by suddenly discovered him/herself as active audience. To date, this practice still prevails in the streets and shopping centers. Between June 29 and July 1, 1962, when President John F. Kennedy and his wife paid an official visit to Mexico City, the local inhabitants made long queues on sidewalks to watch the motorcade. In addition, they were in showcases for a close-up view of the details as broadcasted on TV. This was recorded by Agustín Casasola in photographs from the interior of an appliance store, showing curious passers-by trying to watch the TV set placed in the middle of the showcase [17].

Display cabinets were not the only public spaces where TV sets were exhibited. Canteens, cafes, schools, hospitals, and even some parishes became recurring scenarios, attended by a faithful massive audience, offering far more amusement and comfort than streets. "We met on Saturdays. We gathered in a famous local cafe called *Kikos*, on *Michoacán* street, at *Colonia Condesa*. [...] We gathered to watch freestyle wrestling!," said one interviewee in Mexico City, who in his teens attended weekend meetings regularly [18]. Again, Mayo brothers and the Casasola archives allow tracking back in time these public TV-watching practices. Their photographs depict the daily life of canteens in the city, including the typical small screen turned on in the background [19]. Amidst a male audience, the TV set becomes an essential item that supplements the bar services. The TV set is incorporated into socialization practices, even becoming the spotlight when sports and political events took place. For many TV spectators, this was the only contact at hand with the communication media, and for others, the possibility of breaking with domestic everyday life.

In parallel with the routines of viewers on streets, bars, and restaurants, TV sets were also installed in some hospitals, nursing homes, orphanages, schools, parishes, hotels, offices, and trade unions. Except for the last three cases, TV sets usually came from donations. The appliance was intended to encourage the socialization of hospital patients, as an aid for pedagogical activities of infants, and for the "controlled" recreation of parishioners. The second group of establishments involved commercial interests or the promotion of spaces for rest and leisure of workers and users of certain public places. Occasionally, this practice was a news topic. In May 1956, *TV-56* reported that *Casa Philips* had given a TV set to the cancer ward at the general hospital, in Mexico City: "A group of more than 100 patients were ecstatic watching TV—some of them had never had this experience before" [20].

3.1.2 The private space and the viewer

From its pilot phase in the laboratory, "watching TV" was intended as a household activity, to be enjoyed in the company of family members. The TV set was advertised as a domestic appliance. "A New Era in the history of Mexican houses begins today ...," said *RCA Victor* on September 1, 1950, in the *Excelsior* newspaper [21]. It was not only a matter of advertising, but also of conception. With time, as Silverston points out, television succeeded in being integrated within the "private household culture" [22]. It was a steady and complex process.

The most usual scenario consisted in adapting the living room or a small lounge to place the new appliance. The chairs or the main armchair in the room were arranged around the TV set, so as to facilitate the vision of all viewers ([23], p. 326). "The place was set to ensure satisfactory "TV-watching "sessions; beyond the TV shows displayed, spectators made efforts to accommodate their lifestyle and family gathering spaces" [4]. Spigel compares this "television room" at home with the bourgeois "living room" of Victorian times. The intention in the nineteenth century was to build "true theater halls" for family entertainment [24]. The theater was coming to the household, which then became a theater itself. The sense of this practice is materialized with the arrival of television, says Silverstone. "The living room acquired a new function, or better yet, acquired diverse functions" [4]. In the consumer society of the mid-twentieth century, domestic spaces became [25] multifunctional.

The photographic records of the Mayo brothers and the Casasola archives, as well as printed advertising and cartoons, confirmed the domestic nature of television across virtually all socioeconomic sectors of society [4]. However, testimonies, interviews, and figures confirm that such reality has multiple edges. In Mexico, the idyllic image of a nuclear family sitting in front of a TV set did not materialize in 1950 (**Figure 3**). This image started making sense in the 1960s, with the massive sales of TV sets boosted by lower prices, high manufacturing volumes, greater competition, credit facilities, the consolidation of the television industry, and the strengthening of the middle classes. In the 1950s, the image disseminated by advertising agencies was more a vision derived from the Americanization of popular culture than a palpable common fact. The 1950 population census revealed that 60% of houses in Mexico City had a single room, and 25% had two rooms. Seventy percent of houses were of artisanal construction, made of adobe, wood, stakes,

Figure 3.
Publicity Zenith. In: Novedades, September 7, 1951, Mexico, p. 13.

sticks or stones, and 18% were built of brick and masonry. By the end of the 1940s, only 23% of the population had access to basic potable water services, while 79% of the house occupants were tenants. By the mid-twentieth century, "nearly half of the population lived in tenement houses" [26].

With television, the world, "the outside," different, broad, and ever-changing, entered the house, "the inside," the private and known [27]. It was now possible to explore the outside from the comfort of the living room, with no need to cross the street. The house was the essence of the new communication media. Both industry and viewers sought to make this fact true, although the physical and symbolic spaces of households—and families—would never be the same.

3.1.3 The semi-public space

In the 1950s, watching television was an activity associated to the encounter with others. This collective nature of the activity materialized in at least two scenarios: social gatherings with neighbors, friends, and family, and home businesses that offered TV-watching. Either case involved private spaces, offering no free access without the TV set owner's authorization upon setting a few basic agreements. In any of these circumstances, the private house setting—previously a site of family privacy—suddenly acquired a semi-public character [4].

"I don't have a TV set, as they are very expensive and the credit schemes offered by some shops are unaffordable for a family like ours. So, I watch some TV programs at my sister's, who has TV," replied Guadalupe de Lozano, an inhabitant of *Colonia Portales* in a survey conducted by *TV-56* [28]. This testimony refers to a sporadic "TV-watching" modality. Those who attended these sessions did so under certain restrictions and mediated by third parties. The reception was variable. In theory, this viewer should be more selective and likely more demanding with the TV programs offered. The Mayo brothers archives include numerous photographic series of gatherings with family and friends around the TV set [29]. Children were the ones most benefited from these practices, as they had more flexible schedules and outdoor activities, surrounded by neighbors willing to share their TV set, as evidenced by several of our interviewees [14, 18].

Certainly, routines and tastes of children were changing. The Catholic magazine *Señal* indicated in 1958 [30]: "there is no way to tear children off the screen." This reality had produced a distortion in the infants. Television images were in the minds of the little ones during their hours of study, sport, play, or learning. The new invention ended up keeping minors away from the company of their parents and their educational, religious, and domestic responsibilities. Spaces such as lunch or dinner had also been altered: "[...] The food is served on a special television table so that there is maximum visibility and in semi-darkness," said Guy Robin's article reproduced in *Señal*. From this point of view, the child could not distinguish between reality and screen fiction. The infants acted by imitation. The bad contents of the environment were a danger for the whole society and its future citizens. Additionally, certain programs could cause psychological damage and emotional instability, affect visual health, and lead to a tendency to not exercise [31].

On the other hand, places such as tenements—a multi-family group of houses that became popular in Mexico City—reinforced the collective nature of TV-watching. Specialized magazines such as *Tele-Guia* received letters from readers that were signed by up to 20 neighbors expressing an opinion on the schedule broadcasted [32, 33]. It is quite possible that many of these TV sets were communally owned. From these practices, two elements are confirmed: first, that TV sets favored the gathering of large groups of persons, fostering a collective get-together and entertainment event; and, second, that viewers responded actively to television contents, expressing their own opinions and forwarding them—even collectively—to the press in order to be heard.

These spontaneous practices gave rise to a novel domestic business modality: selling television sessions. In an interview published in 1993, Manolo Fábregas, director and producer, commented that during the transmission of their "teleteatros" he received thank-you letters from persons who had earned extra income from public displays of his programs on their domestic TV set [34]. One of our interviewees recalls that charging "some 50 cents," a neighbor allowed her and other children to watch the afternoon children's TV programs. At times, each attendee carried his/her own chair. Furthermore, she points out that the charge could be higher for watching football and box broadcasts [35].

The Entertainment Office of the General Direction of Interior Affairs of the Federal District acknowledged the existence of "TV rooms" and established that the owners of this "kind of entertainment for profit" should meet the same requirements of movie theaters. In other words, these should have the approval of the General Direction of Public Works, the Police Office, the Fire Department, the Secretariat of National Economy, the Department of Electric Power Control, the Secretariat of Health and Welfare, and the Department of Sanitary Engineering, in addition to a certificate of no debit of fines [36]. This type of standard leads us to deduce that there were public TV projection rooms that were more sophisticated than domestic living rooms, equipped with more infrastructure to provide the service. For its part, the informality of neighbors responded more to the spontaneity and inventiveness than to the logic of a stable business that would allow them to meet all the requirements established by the Direction of Interior Affairs. On the other hand, the business modalities that emerged to meet the "TV-watching" demand were broadened further to include TV-set rental. This activity is traceable through classified ads. The *Compañía Panamericana*, in addition to renting typewriters, compressors, and cars, offered TV sets for rental for a limited time with home-delivery service [37]. Publicity advertised the activity as the seasons' great novelty [37].

The street, coffee, or hospital viewer, the tenement spectator who attended improvised public rooms, watched TV at home or at occasional family gatherings, all lived an authentic TV-watching experience, mediated by the need to share a space, the willingness to make common a personal good, and the fascination with the new technology. A TV viewer in Mexico City was being recognized as a community member. Given the inherent material, socioeconomic, and cultural conditions associated to the context, this collective nature of watching television involved a thin boundary between the public and private. Some previously intimate spaces opened their doors. Sometimes, this was the result of genuine solidarity, whereas in others there was an economic interest involved. *Novelty* was the core that motivated the development of complex social relationships, negotiations, and interaction rules and practices, with both the other spectators and with the channel and messages broadcasted. The high cost of the first TV sets led to recursive practices and community practices to facilitate access to the appliance. Of course, all these experiences clashed with the *domesticity* image that TV-set manufacturers and their advertising campaigns sought to communicate to consumers. Public and semi-public "TV-watching" practices coexisted with that private, homely, and intimate ideal of being in front of the small screen [4].

4. Technological convergence in television reception

The 1950s are unprecedented for the media landscape of Mexico. For the first time, three audiovisual media converge within the same time and space: the trajectory and dynamism of the film industry, the expansion and consolidation of the radio, and the expectation and innovation of television. We refer to two media already well established and one in full materialization. The period represents the capacity to multiply audiences, and with them the meanings of program contents and their penetration at a large scale. Television inaugurated a context of technological convergence. Its incursion into the communications platform represented a novel alternative of information and entertainment, which combined operating mechanisms and offer of services already popularized by other media. Castells points out that network of devices led to the materialization of a mass society and culture ([38], p. 363). The new media, its messages and viewers had to respond to this context of convergence and sociocultural changes. It is worth remembering that in the early 1950s, Mexico had 25,791,097 inhabitants [39]; 10 years later, the General Population Census recorded a total of 34,923,129 persons, almost 18 million of whom lived in cities, equivalent to a little more than 50% of the country's population [40]. Urbanization was asymmetrical. Forty percent of the urban population was concentrated in the capital of the country ([41], p. 700). At the same time, the buoyant middle class experienced a significant rise in Mexico.

The 1950s are a sort of turning point between the predominantly rural Mexico and the Mexico dominated by urban majorities. The period witnesses the coexistence of modernization as an imperative and distrust of the new as a sign of uncertainty. Technological convergence experienced both the suspicion and skepticism of some and the euphoria of others. Not all social stakeholders were prepared. The "great public" reacted from their previous experience and understood the peculiarities of television in function of what their impressions already acquired with other communication media. Its referent was the moving image provided by cinema and the immediacy of information and domesticity brought by the radio.

4.1 The known

The cinema arrived in Mexico in a moment of "faith in progress" and urban renewal, in 1896 [42]. The expansion of the show led to the construction of special theaters and projector tours to various parts of the country. Quickly, Mexico went from being a recipient country to a country where film productions were made. The first experiments of Mexican entrepreneurs and directors, with "The Mexican Charros," a short film of 1903, evolved into the legendary "Santa" in 1931, credited for being the first sound production in the country. The growth of the media consolidated a profitable industry. Between 1930 and the 1950s, the national cinema experienced a "golden age" [43, 44]. As an alternative to the films arriving from Hollywood, this period was characterized by a nationalist cinema that repeatedly alluded to revolution times [4]. By the mid-1940s, around 50% of the films exhibited in Mexican theaters were domestic productions. In spite of this, and "except for the work of Buñuel," the success of the nationalist scheme and its thriving industry went into decline by the mid-1950s ([45], pp. 521-523).

On the other hand, the first public radio broadcasts in Mexico took place in 1921 [46]. Radio stations were founded in 1923, mandated by the federal government, licensed to CYL, owned by Raúl Azcarraga, and to CYB, owned by the cigarette company *El Buen Tono*. Until then, the country's broadcasting system had been managed by the state, so that licensing to private enterprises gave way to a mixed model regulated by the Law of Electrical Communications in 1926 [30]. The idea of turning the radio on "mass entertaining and business enterprise" boosted its popularization, and the radio set entered households to stay [46]. With the expansion and modernization of radio stations, Azcárraga founded XEW in 1930. By 1934, Mexico had 57 radio chains, which further increased to 100 by 1940 ([47], p. 639).

The FM frequency arrived at the end of that decade, but the regulations were not amended until 1961, with the Federal Radio and Television.

4.2 The novelty

The birth of television in Mexico is linked to the fascination with technological innovation. Such interest was linked to the findings of Guillermo Gonzalez Camarena, who in 1946 founded the first experimental television laboratory in the country. Consequently, on September 7 of this year, a small television station under the name XEIGC was installed in 74 Havre street, in Mexico City. This experimental channel broadcasted its own programs every Saturday. Gonzalez had been conducting research since 1935, allowing him to develop a chromatic image system that was tested and presented to the public, from his home, in 1939 [48]. The invention was patented as a tri-chromatic system, based on the use of primary colors for image capture and reproduction [48].

The *Radiolandia* magazine captured the expectation produced by the test emission of González in 1949: "We were able to admire the work by Camarena until recently, when he installed a small TV set in some cinemas in the capital; this was a real surprise for many, who started going to the cinema not to watch the movie, but to be able to observe something they believed was impossible: television in Mexico" [49]. With González Camarena, a first collective seduction by technological advancement emerged. "Mexico is probably the first country having TV with natural colors that, as Gonzalez stresses, is the true television [49]." His inventiveness was a source of pride for Mexicans. The "nationalist consensus" referred to by Loaeza was evidenced in the frequent exaltation to the Mexican as being unique and authentic: quite an example to follow ([50], p. 133). Technological leadership became a patriotic reference, which celebrated with honors that Mexico was the first Latin American country where television was launched.

4.3 Confluences

In technological convergence, television and TV spectators fulfilled two functions in the 1950s: first, *being the novelty*, an object of wonder and curiosity. "And how are we going to be able to watch people in a box?" recalled one of our interviewees as an example of her stupor as a child when experiencing the new technology [51]. And second, *being the synthesis*. Television was presented as a "hybrid" between movies and radio. The best of both worlds was now contained in one single option. Television resembled sound films, offering the appealing benefit of not requiring a public setting for projection. And as the heiress to the radio, it based its operation on a domestic receiver set able to deliver distance and real-time broadcasting.

What was television like? A few months before the inauguration of the television system in Mexico, specialized magazines, such as *Club 16 mm* and *Radiolandia*, conducted pedagogy work on the media and its technique. These magazines published graphic displays showing the mechanisms of cathode ray tubes and made readers become acquainted with the creation of antennas. Referring to the Decree issued on February 11, 1950, *Club 16 mm* specified the technical and administrative standards regarding the installation of television stations. Only one cinema magazine and one radio magazine could provide updated illustrations on the new media. How to make television? At the operational level, interaction was organic. The earliest technicians, cameramen, illuminators, and scenographers—not to mention artists, scriptwriters, announcers, directors, and producers—came from the radio and cinema. Their mission was to readapt the technical expertise acquired in other media. Scenographies and studios started being shared. At the end of 1950, Channel 2 employees received the first training course on television in Mexico, led by the engineer Roberto Kenny and endorsed by the Columbia College of Chicago [52]. "The industry was breathing activity, dynamism and youth," stated Miko Viya, one of the TV pioneer directors [48].

In the 1950s, "making television" and "watching TV" involved a variety of adaptations. While the new medium structured and defined itself, it also modeled the viewers. The processes were two-way. The audiences did not take long to get an impression of "the newcomer" and identify themselves as users. Technological convergence assumed a leading role in this process. The experience gained with other audiovisual devices was key to function, as well as to establish differences and similarities.

5. Conclusions

Technological convergence is neither a recent experience nor a subject restrained to the digital world. In broad terms and without forcing anachronisms, it is reasonable to assert that television represented, in the communications field, the first example of interconnection of services, innovation, and technical process in a single device. In the mid-twentieth century, the world is in an "early convergence" of content and technology.

In this analytical framework, the viewers of the 1950s were key stakeholders of their time. Technological transformation, which for the first time brought together cinema, radio, and television in the same space, was the first reality that viewers had to face. In countries like Mexico, the viewers' practices responded to their impossibility to buying TV sets on a massive scale, to the sense of collectiveness, to social mobility aspirations, to cultural references, and to a context involving growing urbanization, literacy, mass culture, and modernization.

From the moment when TV screens were turned on, the television audience became a plural and changing agent. The material, social, and cultural conditions of the time determined what, how, when, and where to watch TV. This research confirmed that, despite of its recent arrival, audiences in Mexico were active for the most part, and although a certain conservative tone prevailed, they were in constant dialog with the contents projected. Previous experience with cinema and radio, as well as the trajectory of the press, served as a training ground to this end.

Technological convergence, from a historical point of view, constitutes a context, a succession of transformations, and a set of assimilation, integration, and interaction practices. In the mid-twentieth century, television succeeded in linking the known with the novelty, offering speed, connectivity, domesticity, and massification of messages. In an unprecedented way, the TV set took remote motion images and sound to households. It connected the "indoors" with the "outdoors." The effects were multifold. According to McLuhan, the rise and confluence of media were key for remodeling and restructuring the patterns of social interdependence and private life in this era ([53], p. 8). That is why the future of technological convergence and television lies also in the historical reflection, the recognition of the past, and the signs of change that these represent.

Acknowledgements

This article was published due to the financial support of the Universidad Iberoamericana.

Author details

Laura Camila Ramírez Bonilla
Universidad Iberoamericana, El Colegio de México, México

*Address all correspondence to: lcramirezb@gmail.com

IntechOpen

References

[1] Orozco G. Televisión, audiencias y educación. Mexico: Gedisa; 2001

[2] Castells M. Comunicación y poder. Barcelona: Siglo XXI; 2012

[3] Huertas A. La audiencia investigada. Barcelona: Gedisa; 2002

[4] Ramírez L. Moralización y catolicismo al arribo de la televisión. Bogotá y Ciudad de México [Thesis]. El Colegio de México: Mexico; 2017

[5] Castellot G. La televisión en México. Mexico: Adamex; 1999

[6] Bohmann K. Medios de comunicación y sistemas de información en México. Mexico: Consejo Nacional para la Cultura y las Artes; 1989

[7] TV-56. "El televidente opina". Mexico, November 20, 1956. p. 24

[8] Los publicistas elogian y apoyan el esfuerzo de Televisión de México S.A. Novedades, México, 30 de agosto de 1950

[9] UNESCO, La Télévision dans le monde. Rapport sur les moyens techniques de l'information. París: UN-UNESCO; 1954

[10] Mejía F. La industria de la radio y la televisión y la política del Estado mexicano (1920-1960). Fundación Manuel Buendía: México; 1989

[11] UNESCO. Statistical reports and studies. Statistics on radio and television 1950-1960. París: UNESCO; 1963

[12] Archive of Mayo Brothers. Photo library of Archivo General de la Nación (Mexico's General Document Archives). Folder: HMCN 2347-3. Set: "Tele (Aparatos y Gente)". Without date

[13] Novedades. Advertisement of "El Palacio del Hierro". Mexico. August 13, 1950

[14] Interview to ELB (Woman, 1942). Mexico. May 11, 2015. By Laura Ramírez

[15] El Nacional. Mexico, September 13, 1950

[16] McQuail D. La acción de los medios. Los medios de comunicación y el interés público. Amorrortu editores: Buenos Aires; 1998

[17] Casasola Archive. Visit of J. Kennedy to Mexico. National Photo Library INAH. Inventory Number: 19078. June 1962

[18] Interview with FZ, November 19, 2015, Mexico City. By Laura Ramírez B

[19] Archive of Mayo Brothers. Photo library of Archivo General de la Nación (Mexico's General Document Archives). Folder: HMCN 2347-3. Set: "Tele (Aparatos y Gente)". Without date

[20] TV-56. "La casa Philips donó un magnífico aparato de televisión para los enfermos cancerosos del Hospital General". Mexico. May 2, 1956. p. 20

[21] Excélsior. Publicity RCA Víctor. Mexico. September 1, 1950. p. 11

[22] Silverstone R. Televisión y vida cotidiana. Amorrortu editores: Buenos Aires; 1994

[23] Ramírez L. La hora de la TV: la incursión de la televisión y telenovela en la vida cotidiana de la Ciudad de México (1958-1966). Historia Mexicana. 2015;**257**:289-356

[24] Spigel L. Make Room for TV: Television and the Family Ideal in Post-War America. Chicago: University of Chicago Press; 1992

[25] Ibañez J. Por una sociología de la vida cotidiana. Madrid: Siglo XXI; 2014

[26] Greaves C. El México contemporáneo. In: Escalante P, editor. La vida cotidiana en México. Mexico: El Colegio de México; 2010. pp. 241-278

[27] Monsiváis C. Lo entretenido y lo aburrido. La televisión y las tablas de la ley. In: Monsiváis C, editor. Aires de Familia, cultura y sociedad en América Latina. Mexico: El Colegio de México; 2000. pp. 211-254

[28] TV-56. El televidente opina. México. November 20, 1956. p. 24

[29] Archive of Mayo Brothers. Photo library of Archivo General de la Nación (Mexico's General Document Archives). Folder: 2347,2-A. Set: "Venta de TV en funcionamiento, gente viendo la TV". Second Part

[30] Señal. "¿Qué tanto influye la "tele" en el carácter de los niños?". Mexico. August 17, de 1958. p. 10

[31] Ramírez L. "¿Qué niño se resiste a la tele?" Moralidad y prácticas de los infantes ante el surgimiento de la televisión en la ciudad de México (1950-1962). Trashumante. 2016;(8):226-252

[32] Tele-Guia. Letter sent by "los inquilinos de Chimalpopoca 83" to the section "Páginas del director". México. October 1-9, 1963

[33] Tele-Guia. "Páginas del director". Mexico. October 1-9, 1963

[34] Interview with Manolo Fábregas by Laura Castellot de Ballin, Castellot L. Historia de la televisión en México narrada por sus protagonistas. Mexico: Alpe; 1993

[35] Interview with ROP (Male, 1943). December 11, 2016. Mexico City. By Laura Camila Ramírez B

[36] Gaceta Oficial. Department of the Federal District. Notice from the General Directorate of the Interior. Entertainment Office. Mexico. January 20, 1954. p. 1

[37] El Universal. Publicity "Alquilo Televisores". México. August 24, 1951

[38] Castells M. La era de la información. La sociedad red. Barcelona: Siglo XXI; 2008

[39] INEGI señala la cifra exacta de 32.347.698 habitantes para ese año. Ver: INEGI, Anuario estadístico; 1958-1959. p. 35

[40] INEGI, Anuario estadístico 1970-1971. p. 29-30

[41] Club 16 mm. Nuestra televisión. Mexico. March – April, 1950

[42] De la Torre J. La Ciudad de México en los albores del siglo XX. De los Reyes A. Historia de la vida cotidiana en México: siglo XX, la imagen, ¿espejo de la vida?. México: El Colegio de México; 2006, pp. 11-45

[43] De los Reyes A. Medio siglo de cine mexicano (1896-1947). México: Trillas. p. 1987

[44] Silva J. La Época de Oro del cine mexicano: la colonización de un imaginario social culturales. Culturales. 2011;**13**(VII):7-30. http://www.redalyc.org/pdf/694/69418365002.pdf

[45] Standish P. Desarrollo del cine mexicano. Actas XLII (AEPE). Centro Virtual Cervantes. pp. 519-528. http://cvc.cervantes.es/ensenanza/biblioteca_ele/aepe/pdf/congreso_43/congreso_43_64.pdf

[46] Ornelas R. Radio y cotidianidad en México (1900 – 1930). De los Reyes A. Historia de la vida cotidiana en México: siglo XX, la imagen, ¿espejo de la vida?. México: El Colegio de México; 2006. pp. 127-169

[47] Loyo E, Aboites L. La construcción del nuevo Estado, 1920-1945. Nueva

Historia General de México. México: El
Colegio de México; 2010. pp. 595-651

[48] Septién J. La industria de la radio y
la televisión en México. México: Cámara
Nacional de la Industria de Radio y
Televisión; 1991

[49] Radiolandia. Evolución de la
televisión en Mexico, December 2, 1949

[50] Loaeza S. Clases medias y política en
México. El Colegio de México: Mexico;
1988. p. 12

[51] Interview with RMT (Woman, 194-).
November 15, 2012, Mexico City. By
Laura Camila Ramírez B

[52] González G. Historia de la
televisión mexicana (1950-1985).
México: Agrupación de Iniciadores de la
Televisión; 1989

[53] McLuchan M. El medio es el masaje,
un inventario de efectos. Buenos Aires:
Editorial Paidos; 1969

Chapter 3

High-Efficient Video Transmission for HDTV Broadcasting

Yasser Ali Ismail

Abstract

Before broadcasting a video signal, redundant data should be removed from the transmitted video signal. This redundancy operation can be performed using many video coding standards such as H.264/Advanced Video Coding (AVC) and H.265/High-Efficient Video Coding (HEVC) standards. Although both standards produce a great video resolution, too much data are considered to be still redundant. The most exhaustive process in video encoding process is the Motion Estimation (ME) process. The more the resolution of the transmitted video signal, the more the video data to be fetched from the main memory. This will increase the required memory access time for performing the Motion Estimation process. In This chapter, a smart ME coprocessor architecture, which greatly reduces the memory access time, is presented. Data reuse algorithm is used to minimize the memory access time. The discussed coprocessor effectively reuses the data of the search area to minimize the overall memory access time (I/O memory bandwidth) while fully using all resources and hardware. This would speed up the video broadcasting process. For a search range of 32 × 32 and block size of 16 × 16, the architecture can perform Motion Estimation for 30 fps of HDTV video and easily outperforms many fast full-search architectures.

Keywords: HDTV broadcasting, motion estimation, H.264/AVC, H.265/HEVC, video coding, video transmission

1. Introduction

Broadcasting is the distribution of audio or video content to a dispersed audience via any electronic mass communication medium but typically is the one using the electromagnetic spectrum (radio waves). Recently, broadcasting operations are not used only by television signal, but also it includes many smart devices such as cell phones, video phones, and video conferencing. Two main problems were highlighted because of the high demand of video applications that need broadcasting [1–3]. The first problem is the huge bandwidth needed for transmitting such huge video data. The second problem is the delay in the video transmission process due to the huge computations required for video coding and transmission process [4, 5]. Many video coding standards tried to solve such problems. H.264/Advanced Video Coding (AVC) and H.265/High-Efficient Video Coding (HEVC) are the most recent video coding standards that tried to tackle the aforementioned problems [6, 7]. Although the two video coding standards provide a great video resolution and a low bit rate (BR), the computational complexity required for the video encoding process is very high [1, 3, 8]. Consequently, the video transmission speed will slow down. As a result, using H.264/AVC and/or H.265/HEVC standards may not be suitable for real-time video broadcasting applications.

H.264/AVC [9] is a video coding standard that was developed by ITU and ISO [10]. The main advantage of H.264/AVC standard is to minimize the bit rate of the transmitted video signal. It achieves up to 50% savings in the transmitted video bit rate if compared to other previous standards (please see **Figure 1**). As a result, H.264/AVC standard is used in many video applications such as HD-DVD, multimedia streaming, remote video surveillance, medical image processing field, video conferencing, HDTV broadcasting, video on demand, and multimedia messaging [5, 6, 11]. Multiple reference frames, half-pel and quarter-pel accurate Motion Estimation, using small block size, exact-match transform, adaptive in-loop deblocking filter, and enhanced entropy coding methods, and variable block size techniques are used to achieve low bit rate and high resolution of the transmitted video signal [12].

Due to the high demand of high-resolution video applications and the traffic constrains of the network infrastructure, the offered transmission bit rate of H.264/AVC standard is not suitable to fulfill such application needs. ITU-T VCEG and ISO/IEC MPEG [13] developed a new video coding standard called H.265/High-Efficient Video Coding (HEVC) standard. H.265/HEVC standard was developed for three main goals. The first goal is to be able to encode high-resolution video sequences such as 4 K and ultrahigh definition (UHD). The second goal is to lower the video transmission bit rate by approximately 50% compared to the H.264/AVC standard. The last goal is to speed up the video coding and transmission process by utilizing parallel processing operations. It is worth mentioning that H.265/HEVC encoder is four times more complex than older standards [1, 14]. However, from the previous aforementioned discussions, it is clear that both H.264/AVC and H.265/HEVC standards require processing too much data in order to obtain higher compression and accordingly low bit rate. This is why the implementation of both standards is not that easy because of the high demand of huge memory size that is being able to process such huge data. This is the main bottleneck in such video coding standards,

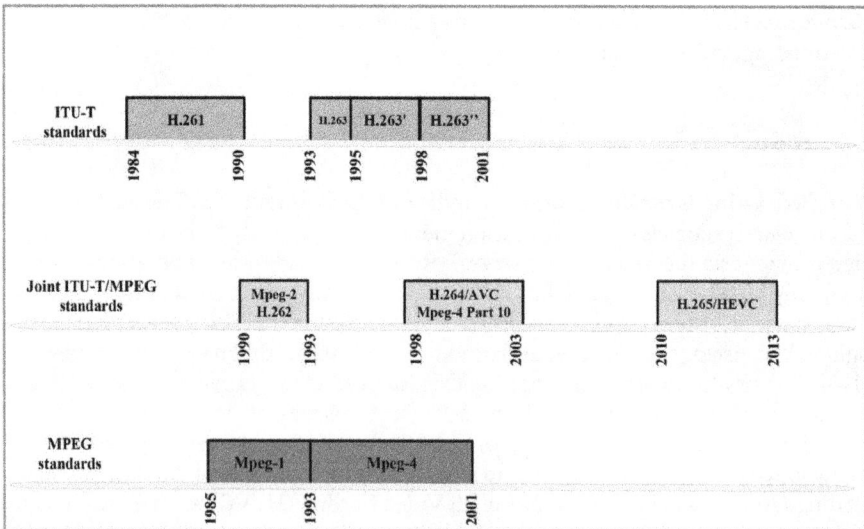

Figure 1.
Video coding standards.

since fetching such huge data from memory will increase the memory access time and slow down the encoding process. As a result, those video coding standards may not be suitable for real-time video applications such as HDTV broadcasting.

The huge data and memory access time that are needed for HDTV broadcasting, if using either H.264/AVC or H.265/HEVC standards, are the recent problems under study for many researchers nowadays. Some techniques tried to reduce the search area size in the Motion Estimation process in order to reduce the memory access time. Adaptive search window size (ASWS) technique [15, 16] adaptively decides the size of the search area according to the motion activity of the Current Block. The authors in [15] used three window sizes 3×3, 7×7, and 15×15. The main drawback of such algorithm is the lake of video visual quality since the selected window sizes are not enough to cover fast motion activities of certain Current Blocks. In [2], this problem is avoided by adaptively adjusting the search window size according to some model equations that used some parameters. Those parameters changed according the motion activity of a Current Block and considering the motion activity of surrounding blocks. The accuracy to decide the search window size in [2] is very high, and, accordingly, the memory access time is less than the case of using the conventional full-search Motion Estimation (FSME) process. Although the previous techniques greatly reduce the memory access time, their very-large-scale integration (VLSI) implementation is not easy [17]. They are not suitable for hardware implementation as they lose the regularity of data flow, but they can greatly reduce the computational complexity. Some straightforward VLSI architectures are used to implement the FSME of the video encoder (either for H.264/AVC or H.265/HEVC standards). Such implementation has many advantages such as they are greatly reducing the memory access time. Additionally, the data flow of their architecture is uniform. This gives such architectures simplicity in their design. The architecture given in [4] is a good example for such architectures. The authors in [4] used a smart algorithm with a simple local memory to reuse data (data reuse level A and level B) of the search area. It means that data could be fetched only once from the main memory. This greatly decreases the memory access time required for the Motion Estimation process. The design proposed in [4] is a flexible one, at which it can be used either for H.264/AVC or H.265/HEVC standards with slightly modifications in the hardware.

The chapter is organized as follows. Section 2 illustrates the problem formulation in details. Section 3 describes the memory I/O bandwidth reduction techniques. Motion Estimation co-processor with data reuse is described in details in Section 4.

2. Problem formulation

The most exhaustive part in video encoding process is the Motion Estimation process [18]. As seen in **Figure 2**, Motion Estimation process consumes up to 53 and 70% of the entire encoding process in case of using one and four reference frames, respectively. This is due to the huge data and computations that are required to perform such operation. As a result, we emphasize on reducing both the computations and the data fetched from the main memory in order to speed up the video encoding process in order to be available for real-time HDTV broadcasting application. In Motion Estimation process, the current frame ψ should be divided into blocks called Current Blocks (please see **Figure 3**). The size of each block is M × M pixels. The main idea of the Motion Estimation process is to reduce the temporal redundancy in the transmitted video sequences. This can be done by searching for a best match candidate block of each Current Block within a search area in the reference frame

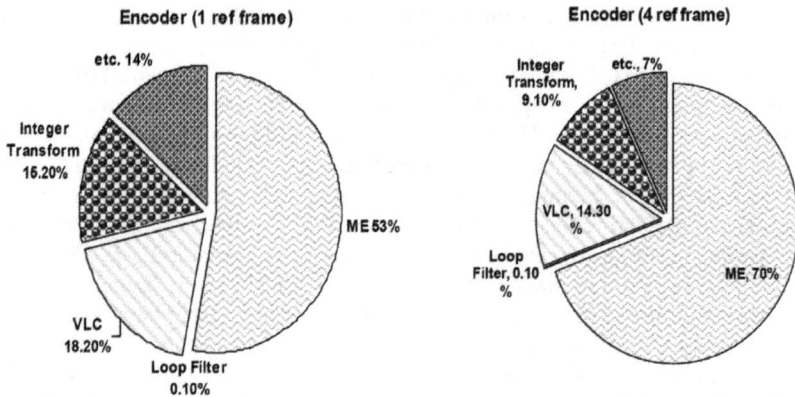

Figure 2.
Motion estimation computational complexity using one and four reference frames, respectively, and using H.264.AVC standard.

$\psi - 1$ as seen in **Figure 3**. Referring to **Figure 3**, the best match candidate block can be calculated by exhaustively searching all candidate blocks allocated in every pixel in the search area. The search area is allocated at frame $\psi - 1$. The search area size is $2X_{max} \times 2X_{max}$, where $2X_{max}$ is the range of the selected search area. The sum of absolute difference (SAD) metric is used to search for the best match candidate block. The pixel in the search area corresponding to the minimum SAD value (i.e., point (k,l) in **Figure 3**) represents the best match candidate block. Two outputs of the Motion Estimation process are seen in **Figure 3**. The first output is the residue between the Current Block and the best match candidate block. The second output is the displacement between the center of the search area and the best match candidate block, represented by the motion vector (MV). The MV can be represented as

$$MV(x,y) \;=\; \arg \min \{SAD(x,y;k,l)\} \qquad (1)$$

where $-X_{max} \leq k,l \leq X_{max}$, x, and y are the center coordinates of the search area and k and l are the coordinates of the best match candidate block.

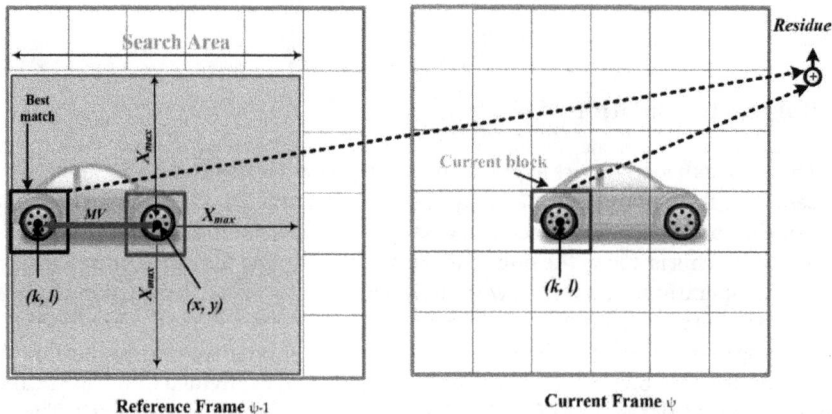

Figure 3.
The current and the reference frames in ME process.

	Frame size (pixels)	Frame rate (frames/sec)
HDTV broadcast	1920 × 1080	30
SDTV broadcast (D1)	720 × 486	30
Digital cinema (DC)	4096 × 2160	24 fps
Standard definition (SD)	720 × 486	30 fps
Video conferencing (SIF)	352 × 240	30
Internet streaming video (QSIF to SIF)	176 × 144 352 × 240	30
Desktop video phone (QSIF)	176 × 144	15–30

Table 1.
Video encoding formats [4].

It is worth mentioning that video applications consume much data compared to some other multimedia sources such as image, speech, and text. The number of bits required to transmit video sequences depends on the video frame resolution. The lager the frame resolution is, the larger the number of bits required to be transmitted. Consequently, the transmission bit rate will also increase. Different video frame formats are shown in **Table 1**. Given that the number of bits/pixels required for specific video resolution is B, the number of pixel per line of one video frame is P, the number of lines per video frame is L, and the transmitted frame per second for such video resolution is F, the video transmission bit rate (BR) can be calculated as in Eq. (2):

$$BR = B \times P \times L \times F \qquad (2)$$

It is noticed from **Table 1** that frame size is increased due to the increase of consumer demand for higher resolution [4]. The more the video frame size is, the more the required search area size will be. This increases the data to be fetched from the memory. Accordingly, more memory I/O bandwidth is required. This is a big problem since the memory I/O bandwidth is limited. As an example, HDTV broadcasting requires much data to be fetched from the memory than the Internet streaming video which uses QSIF or SIF video formats. Many recent researches are directed to create different techniques for a better use of the available memory I/O bandwidth. In the following section, a review of most recent techniques that have been used for reducing the I/O bandwidth problem will be performed.

3. Memory I/O bandwidth reduction

The memory I/O bandwidth is the main problem in video encoding process. The more the data is needed from the memory, the more the required memory I/O bandwidth is. As a result, the delay of the video encoding process will be high. This will prevent the use of such video encoding process in real-time applications such as HDTV and SDTV broadcasting. To avoid such problem, the search area data required for the Motion Estimation process in a video encoder should be reused. In other words, the required data for the Motion Estimation process should be fetched only once from the main memory. This will speed up the video encoding process and allow the video encoder to be used in real-time video applications.

During the full-search Motion Estimation, the current frame is divided into nonoverlapped blocks. Each block (Current Block) has a size of M × M pixels. The selected search area has a size of $SA_x × SA_y$. The size of the Current Block and the search area is different according to the frame size. As an example, the Current Block and search area sizes for SIF video sequences are 16 × 16 and 32 × 32, while for SDTV and HDTV are 32 × 32 and 64 × 64, respectively. The search area can be divided into $\{SA_x - M + 1\} × \{SA_y - M + 1\}$ candidate blocks or search pixels. It is worth mentioning that the candidate blocks for the contour pixels use additional padding (please see dashed line area in **Figure 4**). The padding pixels will have maximum values to be sure that they will be excluded from our selections.

Considering the candidate blocks #1 and #2 in **Figure 4**, there will be M − 1 overlapped pixels that should be fetched twice from the main memory. If we continue horizontally, there will be a complete strip that contains several overlapped areas between the candidate blocks of the search area. The horizontal strip of overlapped Data Pixels can be reused and fetched only once from the main memory. In this case, such data is called data reuse level A. If the Motion Estimation process is proceeded by going one step down, huge Data Pixels will be overlapped. This is called data reuse level B, if fetched only once from the memory. It is worth mentioning that a smart architecture of the Motion Estimation processor is required in order to be able to reuse data and reduce the I/O memory access bandwidth. Given that the size of the Current Block is M × M, the number of Data Pixels (DP) required from the memory to perform the Motion Estimation process for only one Current Block is given by Eq. (3). There are much redundant (repeated) data that are fetched from the main memory. This increases the I/O memory bandwidth which is not practical for real-time applications such as HDTV broadcasting:

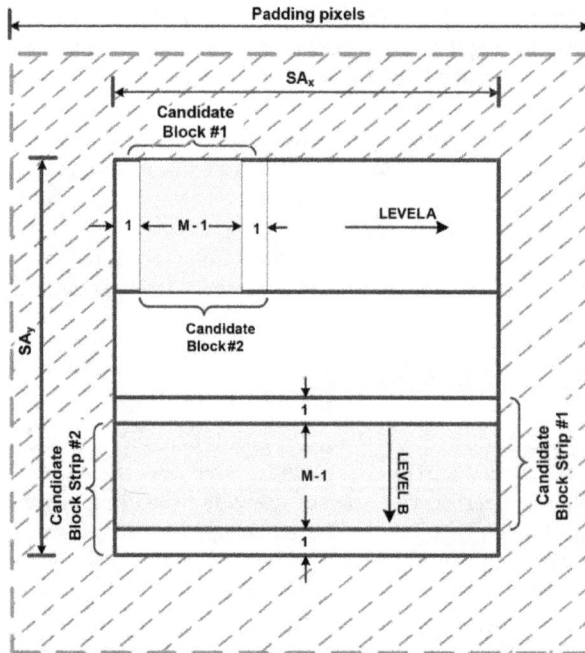

Figure 4.
Data reuse (levels A and B).

$$DP = \{SA_x - M + 1\} \times \{SA_y - M + 1\} \times (M \times M) \qquad (3)$$

Data reuse is the most recent efficient algorithm that is used to reduce the I/O memory bandwidth while performing Motion Estimation process [4, 19, 20]. Many redundant memory access times are skipped during the Motion Estimation process while performing the data reuse algorithm. The algorithm skips the calling of the same pixels multiple times and accesses the required Data Pixels only once while performing the Motion Estimation process. This is how it reduces accessing the main memory many times, and, consequently, it reduces the I/O memory bandwidth. For a maximum reduction of the I/O memory bandwidth, there are four levels of data reuse: levels A, B, C, and D [4, 19]. The four levels are described in the following subsections.

3.1 Data reuse level A

For a horizontal strip as seen in **Figure 4**, there will be an overlapped Data Pixels between any two consecutive blocks. The overlapped area size is $\{SA_y - M + 1\} \times \{M - 1\}$. Data reuse level A principle is achieved by loading only new data from the memory and ignoring old data that are already fetched before from the main memory [4, 19]. This requires a smart local memory in the Motion Estimation processor to keep reusing the old fetched data. As an example, seen in **Figure 4**, when performing the SAD operation between the Current Block and candidate block #2, only M column of Data Pixels will be fetched from the memory. This is a huge saving in the memory I/O bandwidth.

3.2 Data reuse level B

As seen in **Figure 4**, there is another vertical level of overlapped data if the SAD operation is performed one pixel in the down direction [4, 19]. The size of the overlapped Data Pixels is $SA_y \times [M - 1]$. The overlapped Data Pixels will be while performing the SAD operation between the Current Block and both of candidate blocks strip #1 and #2 as seen in **Figure 4**. Data reuse level B is achieved by loading only SA_x Data Pixels when moving from one horizontal strip to the next strip below it.

3.3 Data reuse level C

For two horizontal consecutive search areas $SA1_x \times SA_y$ and $SA2_x \times SA_y$, there will be an overlapped area of a size $\{|SA1_x - SA2_x|\} \times SA_y$ as seen in **Figure 5**. These overlapped Data Pixels could be fetched only once if level C data reuse is applied while performing Motion Estimation process [4, 19].

3.4 Data reuse level D

For two vertical consecutive search areas $SA_x \times SA1_y$ and $SA_x \times SA2y$, there will be an overlapped area of a size $SA_x \times \{|SA1_y - SA2_y|\}$ as seen in **Figure 6**. These overlapped Data Pixels could be fetched only once if level D data reuse is applied while performing Motion Estimation process [4]. Performing data reuse levels C and D require higher complexity in the design of the Motion Estimation process to reuse data already loaded from the former search area strips to the latter search area strips. This is a trade-off between the higher design complexity and the reductions in the memory I/O bandwidth.

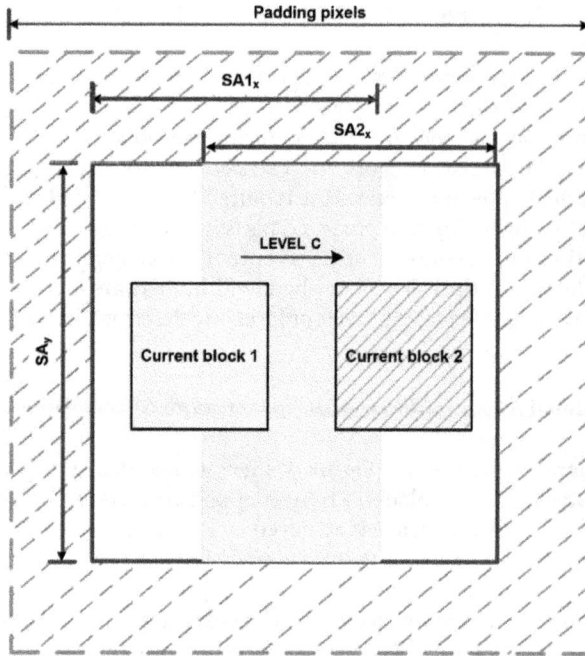

Figure 5.
Level C data reuse.

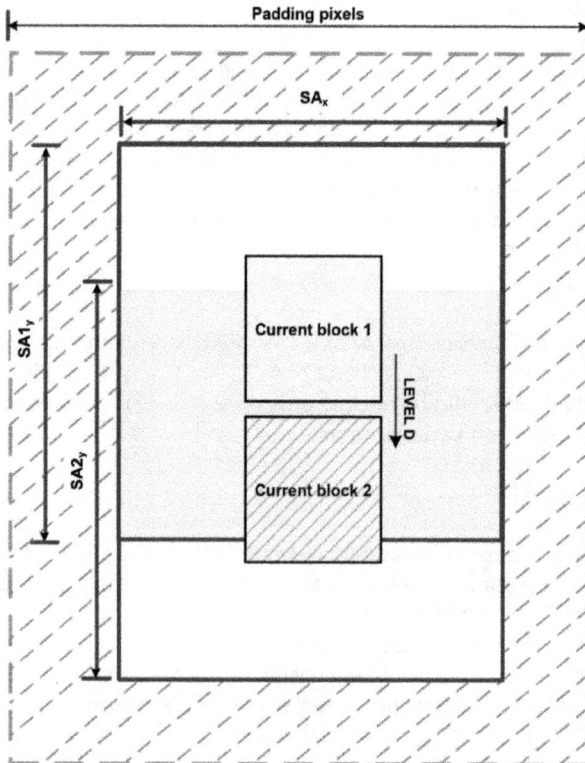

Figure 6.
Level D data reuse.

4. Motion estimation co-processor with data reuse

The main idea that is discussed in this section is how to design a smart Motion Estimation co-processor that could perform data reuse for a maximum reduction in the memory I/O bandwidth. So, such co-processor can be used for real-time video applications such as HDTV broadcasting. The design of the co-processor is preferred to be simple. This requires implementing only level A and level B data reuse since the other two levels require too much hardware complexity. One smart design that is modified and discussed in this chapter is the one proposed in [4]. The modified Motion Estimation co-processor can be used for both H.264/Advanced Video Coding (AVC) and H.265/High-Efficient Video Coding (HEVC) standards. Although the used window size for such co-processor is 32 × 32 pixels with a Current Block size of 16 × 16 pixels, it can be generalized for a higher number of pixels in both the search area and the Current Block to be suitable for HD video sequences. A modified top-level archi-tecture of the proposed Motion Estimation co-processor in [4] is shown in **Figure 7**.

This co-processor is designed for H.264/AVC; however, it could be used for H.265/HEVC standard with a little modification in the hardware. The structure in [4] is designed to perform the Motion Estimation process for a Current Block of size 16 × 16. The search area can be 32 × 32 or more. To increase the speed of the co-processor proposed in [4], the Current Block is directly loaded from the main memory, and the search area will be loaded into a local memory as will be discussed later. The structure consists of a processing element (PE) array. The PE array consists of a 16 × 16 PE that computes the absolute difference (AD) values between the Current Block and a candidate block in the search area. The Current Block will be directly loaded into the PE array from the main memory, while the candidate block will be loaded into the PE array from the local memory. Data reuse levels A and level B are implemented using a local memory. This local memory is very simple and consists only of bank of

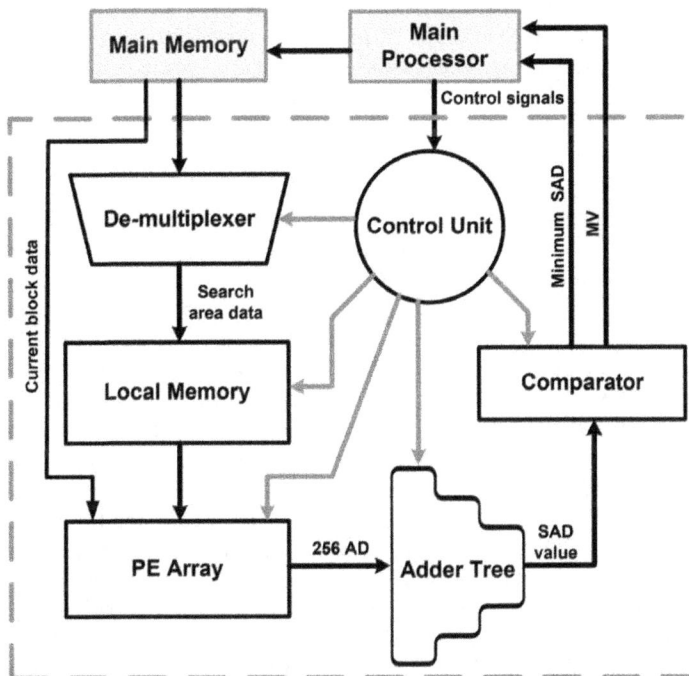

Figure 7.
Motion estimation co-processor.

registers. Using smart control signals, this local memory can succeed to greatly reduce the memory I/O bandwidth. Once the ADs are calculated using the PE array, an adder tree is used to add all of them and calculate a final SAD. The adder tree is adding all values in parallel fashion so that the addition process will be very fast. The SAD will be compared with the calculated SAD so far, and the minimum value will be selected. The candidate block in the search area with the minimum SAD so far will be selected to be the best match candidate block and the corresponding motion vector. The demultiplexer is used to smartly feed the candidate blocks into the PE array.

4.1 Design the local memory

Implementing data reuse levels A, B, C, and D can be performed using a local memory inside the Motion Estimation co-processor. This local memory could fetch the search areas only once from the main memory. This means less memory access time is required. A smart algorithm is needed to control the data flow in/out of the local memory. The local memory in **Figures 7** and **8** is designed to perform level A and level B data reuse. However, it can be used with slight modifications in its hardware and the operating algorithm to perform the other two data reuse levels (i.e., level C and level D data reuse). The modified architecture of the local memory [4] is shown in **Figure 8**. As mentioned before, this co-processor is mainly used for the Motion Estimation process of the H.264/AVC encoder.

The Current Block (CB) is loaded only once per one search area directly from the main memory into the PE array. The 16 × 16 Current Block is loaded into the PE array row by row as seen in **Figure 9**. The CB loading process starts from the bottom PE row and shifted in the up direction as seen in **Figure 9**. Each pixel of the CB is loaded into a register R_x of each PE via pin X_{in} as seen in **Figure 8**. While loading a new row of the CB into the PE array, the content of the register R_X in each PE will be shifted up via the pin X_{out}. This process will be continued until all the CB is loaded into the PE array. Loading the CB into the PE array consumes 16 clock cycles. The local memory is used to load the PE array with the search area of size $SA_x \times SA_y$ (including the padding area). The local memory is very simple in its design. It consists of two banks of registers. Each bank is 16 × 16 byte registers. No addressing modes are required for such local memory. Only a simple 5-bit counter is used for addressing. There is an additional (1 × 2) demultiplexer (Demux) used to control loading the PE array by the search area.

The Motion Estimation process starts by loading the Current Block inside the PE array as mentioned before. The local memory starts loading the search area by loading the top left 16 × 16 candidate block into Register Bank #1 of the local memory as seen in **Figure 8**. This will be performed via connecting terminal 1 of the Demux to the main memory. Loading operation will be performed by loading the candidate block of the search area line by line into Register Bank #1. This will take additional 16 clock cycles.

In clock cycle 33, two operations will be performed at the same time. First, since the first top-left candidate block is available on Register Bank #1 of the local memory, the Motion Estimation co-processor will start loading this candidate block into the PE array column by column starting by the left column of the Register Bank #1 of the local memory. A simple 5-bit counter will be used to read such columns. The pixels of the candidate block will be loaded into the PE via pin Y_{in} and stored into register R_Y (please see **Figure 8**). It is worth mentioning that every time a column of a candidate block is loaded into the PE array, it will be shifted to the left direction via pin Y_{out} (as seen in **Figures 8** and **9**) until the PE array is filled with the candidate block. Second, the second group of the search area will be loaded into the Register Bank #2 of the local memory. This will be performed by connecting terminal 2 of the Demux to the main memory.

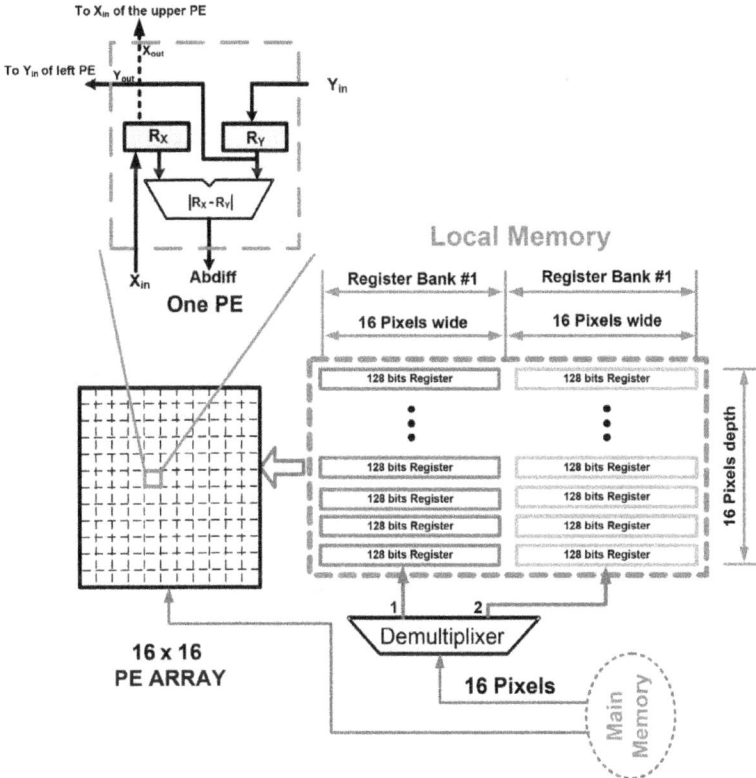

Figure 8.
Local memory design for the motion estimation co-processor.

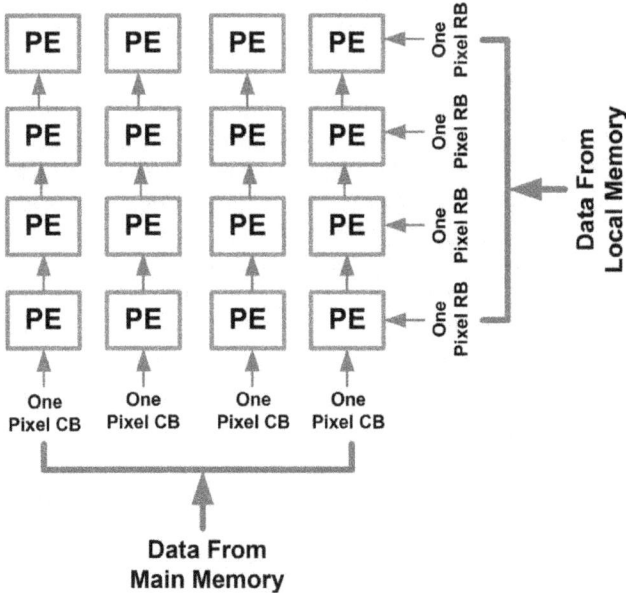

Figure 9.
Loading the current block (CB) and the candidate block (RB) into a 4 × 4 PE array.

Once the candidate block is loaded into the PE array, the PE array will start calculating the AD between each pixel in the Current Block and the corresponding pixel in the candidate block using the adder of each PE. It is worth mentioning that level A data reuse will be performed by shifting only one column from the Register Bank #2 of the local memory into the PE array. While loading the first strip candidate blocks inside the PE array using the Register Bank #2 of the local memory, level B data reuse could be performed by loading new horizontal line into the left bank registers of local memory (Register Bank #1) via terminal 1 of the demultiplexer. Once the last candidate block of the first strip is loaded into the PE array, the counter will back to load a new candidate block (using level B data reuse) by loading the first left column of the Register Bank #1 of the local memory into the PE array. This operation will be continued until all candidate blocks are loaded into the PE array.

Every time a new candidate block is loaded into the PE array, 256 ADs will be calculated by the PEs. The adder tree will calculate the SAD value of the 256 ADs. The comparator will compare the current SAD value with the minimum SAD so far. The final minimum SAD will be decided after all candidate blocks are checked. All of these operations are performed at no stall at all and 100% utilization of the PE array. This is the main advantage of the architecture discussed in this section; this is why such architecture is suggested for real-time high-speed HDTV broadcasting applications. The final output of the co-processor will be the best candidate block with the minimum SAD value and the corresponding MV. These two values will be sent to the main processor to calculate the residue as seen in **Figures** 3 and 7.

5. Conclusion

This book chapter provides a solution for the huge memory access time required by the high-resolution video broadcasting operation (HDTV broadcasting). A high-speed ME architecture that greatly minimizes the memory access time, if adopted into high-resolution H.264/AVC and H.265/HEVC video standards, has been presented. The proposed architecture can perform ME process for 30 fps of HDTV video. It can easily outperforms over many fast full-search architectures due to the great reduction in the memory access time required for the ME process. The proposed architecture effectively uses pipelining, parallelism, and data reuse to achieve a high-throughput rate while maintaining 100% PE utilization. The proposed co-processor is suitable for high-performance and high-accuracy real-time video applications such as HDTV and SDTV broadcastings. As a future work, the author of the book chapter is welling to reduce the processing time required for the ME process. This can be achieved by adopting some dynamic models inside this co-processor to adaptively select the proper size of the search area. The size of the search area is decided using these model equations according to the motion activity of the Current Block. Accordingly, the computations and the memory access time required for the ME process are expected to be greatly reduced.

Acknowledgements

The author acknowledges the support of Southern University and A&M College, Baton Rouge, USA, for its support to finalize this work.

Author details

Yasser Ali Ismail
Electrical Engineering Department, Southern University and A&M College,
Baton Rouge, USA

*Address all correspondence to: yasser_ismail@subr.edu

IntechOpen

References

[1] Jamali M, Coulombe S. Fast HEVC intra mode decision based on RDO cost prediction. IEEE Transactions on Broadcasting. 2018:1-14

[2] Ismail Y, McNeely JB, Shaaban M, Mahmoud H, Bayoumi MA. Fast motion estimation system using dynamic models for H.264/AVC video coding. IEEE Transactions on Circuits and Systems for Video Technology. 2012;22:28-42

[3] Sotelo R, Joskowicz J, Anedda M, Murroni M, Giusto DD. Subjective video quality assessments for 4K UHDTV. In: 2017 IEEE International Symposium on Broadband Multimedia Systems and Broadcasting (BMSB); 2017. pp. 1-6

[4] Ismail Y, El-Medany W, Al-Junaid H, Abdelgawad A. High performance architecture for real-time HDTV broadcasting. Journal of Real-Time Image Processing. 2014;11(4):633-644

[5] Vayalil NC, Kong Y. VLSI architecture of full-search variable-block-size motion estimation for HEVC video encoding. IET Circuits, Devices and Systems. 2017;11:543-548

[6] Rao KR. High efficiency video coding. In: 2016 Signal Processing: Algorithms, Architectures, Arrangements, and Applications (SPA); 2016. pp. 11-11

[7] Hannuksela MM, Yan Y, Huang X, Li H. Overview of the multiview high efficiency video coding (MV-HEVC) standard. In: 2015 IEEE International Conference on Image Processing (ICIP); 2015. pp. 2154-2158

[8] Ismail Y, Elgamel MA, Bayoumi MA. Fast variable padding motion estimation using smart zero motion prejudgment technique for pixel and frequency domains. IEEE Transactions on Circuits and Systems for Video Technology. 2009;19:609-626

[9] Wiegand T, Sullivan GJ, Bjontegaard G, Luthra A. Overview of the H.264/AVC video coding standard. IEEE Transactions on Circuits and Systems for Video Technology. 2003;13:560-576

[10] Advanced video coding for generic audiovisual services. International Telecommunications Union, Telecommun. (ITU-T) and the International Organization for Standardization/International Electrotechnical Commission (ISO/IEC) JTC 1, Recommendation H.264 and ISO/IEC 14 496-10 (MPEG-4) AVC; 2003

[11] Ohm J, Sullivan GJ, Schwarz H, Thiow Keng T, Wiegand T. Comparison of the coding efficiency of video coding standards 2014 including high efficiency video coding (HEVC). IEEE Transactions on Circuits and Systems for Video Technology. 2012;22:1669-1684

[12] Lee TK, Chan YL, Siu WC. Adaptive search range by depth variant decaying weights for HEVC inter texture coding. In: 2017 IEEE International Conference on Multimedia and Expo (ICME); 2017. pp. 1249-1254

[13] ITU-T and JVT. High Efficiency Video Coding (HEVC) Text Specification Draft 6. Joint Video Team of ISO/IEC MPEG and ITU-T VCEG; 2012

[14] Kuang W, Tsang SH, Chan YL, Siu WC. Fast mode decision algorithm for HEVC screen content intra coding. In: 2017 IEEE International Conference on Image Processing (ICIP); 2017. pp. 2473-2477

[15] Goel S, Ismail Y, Bayoumi MA. Adaptive search window size algorithm for fast motion estimation in H.264/AVC standard. In: 48th Midwest Symposium on Circuits and Systems; 2005; Vol. 2. 2005. pp. 1557-1560

[16] Kibeya H, Belghith F, Ayed MAB, Masmoudi N. Adaptive motion estimation search window size for HEVC standard. In: 2016 7th International Conference on Sciences of Electronics, Technologies of Information and Telecommunications (SETIT); 2016. pp. 410-415

[17] Mukherjee R, Banerjee A, Chakrabarti I, Dutta PK, Ray AK. Efficient VLSI design of CAVLC decoder of H.264 for HD videos. In: 2017 7th International Symposium on Embedded Computing and System Design (ISED); 2017. pp. 1-4

[18] Celebi AT, Yavuz S, Celebi A, Urban O. One-dimensional filtering based two-bit transform and its efficient hardware architecture for fast motion estimation. IEEE Transactions on Consumer Electronics. 2017;**63**:377-385

[19] Tuan J-C, Chang T-S, Jen C-W. On the data reuse and memory bandwidth analysis for full-search block-matching VLSI architecture. IEEE Transactions on Circuits and Systems for Video Technology. 2002;**12**:61-72

[20] Goel S, Ismail Y, Devulapalli P, McNeely J, Bayoumi MA. An efficient data reuse motion estimation engine. In: IEEE Workshop on Signal Processing Systems Design and Implementation (SIPS '06); 2006; 2006. pp. 383-386

Section 2

Convergence and Human Aspects

Chapter 4

Culture as a Determinant in Innovation Diffusion

Henrik Vejlgaard

Abstract

This study focuses on how culture may influence the rate of adoption of a television innovation, which is also a service innovation. The starting point is to understand theoretically what consumers may perceive they adopt when they adopt an innovation such as digital terrestrial television (DTT). To understand this, service theory is introduced to help define DTT in the eye of the consumers. The proposition is that a country's culture may affect the rate of adoption of a service innovation no matter how the service innovation is perceived, but if the service innovation is perceived as part of or connected to a service offering that is broadly founded in the culture, there may be a stronger motivation to adopt the service innovation. The study introduces a novel approach to analyse culture in an innovation diffusion context by selecting 10 cultural variables that may have an influence on the rate of adoption of DTT in Denmark. The empirical part of the study concludes that these 10 cultural variables can play a role in innovation diffusion.

Keywords: digital terrestrial television, service innovation, service theory, service package, rate of adoption, cultural variables, national culture, television culture

1. Introduction

One of the big questions in the diffusion of innovations (DOI) research field is what determines the rate of adoption of an innovation. The answer is without doubt complex but research do offer some answers. One answer from literature is that five categories of variables determine the rate of adoption: (I) perceived attributes of innovations, (II) type of innovation-decision, (III) communication channels, (IV) nature of the social system, and (V) extent of change agents' promotion efforts ([1], p. 222). However, these five categories of variables have not received equal attention in research. It is notable that the category that has been most extensively investigated (perceived attributes) has been found to be the most important variable in determining the rate of adoption ([1], p. 222). Historically, DOI research has focused on what determined the rate of adoption of goods and products. But what determine the rate of adoption of products may be different from what determine the rate of adoption of service innovations. This is notable, because, as Dwyer et al. write, 'It seems plausible that intangible service offerings [...] may be more influenced by national culture than are tangible product goods. This area appears especially ripe for exploration' [2]. This study will follow this recommendation and focus on how culture may influence the rate of adoption of an innovation within the broadcasting industry, namely, digital terrestrial television (DTT), with a case study approach. The case is from Denmark.

DTT has been characterised as a digital innovation, specifically 'a new digital television service' [3]. But what does that make of DTT? Is DTT about the adoption of a technological innovation or of a service innovation? A technological innovation and a service innovation are presumably two very different types of innovations. However, there is also a convergence between technology and services, which does not necessarily make it clear if an innovation is about new technology or a new service or both. This chapter will address the convergence of service and technology issue in a diffusion perspective, as this is likely to be one of several convergence issues in broadcasting in the future. For instance, this particular convergence issue may affect the adoptive process of innovations within the broadcasting and other service industries.

The adoptive process of innovations has been studied for more than a hundred years ([1], Chapter 2), resulting in a mainly empirically driven science [4], based on quantitative research ([1], p. 196). While this study is conceptual in nature, it will also follow the empirical, quantitative tradition. Therefore, the aim of this study is two-fold: the first aim is to introduce a novel conceptual approach to analyse culture in an innovation diffusion context (alternative to existing research approaches, cf. the Literature Review). The second aim is to test the conceptual approach empirically.

2. Culture and diffusion of innovations

The word 'culture' has different meanings in different contexts. Often a specifying word is put in front of 'culture' to identify a specific type of culture, for instance, national culture, community culture, or company culture. The early diffusion of innovation research appears to have focused mainly on community culture. This may have to do with the fact that most studies in early diffusion research took place at a community or village level ([1], Chapter 2).

As mentioned in the introduction, there are five categories of variables that appear to determine the rate of adoption. As presented by Rogers, norms and network interconnectedness are two aspects of variable (IV) nature of the social system. Culture may be related to the social system and, indeed, to the other variables as well. However, variable IV appears to relate to the individual's social system or to norms and degree of network interconnectedness at community or village level, not at a national level. Based on this interpretation, national culture is not explicitly included in Roger's overview of variables that determine the rate of adoption of an innovation.

It has been established that national culture is a complex, multifaceted phenomenon [5, 6]. However, it has also been questioned if such a thing as national culture exists [7]. This may have to do with the notion that the term 'national culture' may imply the existence of a uniform and/or normative culture in a country. Some academics argue that cultural variability within nations is so great that it is fruitless to make generalisations with respect to national culture [7]. However, research has found that this objection has little empirical support. National culture does exist in many nations across the globe [8, 9]. This study concurs with Erumban and de Jong who state that 'all individuals live and work within a cultural environment in which certain values, norms, attitudes, and practices are more or less dominant and serve as shared sources of socialisation and social control' [10] and with Dwyer et al. who state that 'culture and country are not the same, we used the country as a surrogate for culture for practical purposes as has largely been done in prior research' [2].

Rogers only makes few references to 'general' culture in *Diffusion of Innovations*. One of the references is to individualistic culture and to collectivistic culture ([1], p. 179), which are both well-documented culture characteristics (cf. [11]). Individualistic cultures are those in which the individual's goals take precedence over the collectivity's goals. Collectivistic cultures are those in which the collectivity's goals take precedence over those of the individual [12]. Each type of culture may have different influence on the adoptive process: in a collective culture, the innovation-decision process may be based on what the community as a whole decides or what community elders decide, whereas in an individualistic culture, the innovation-decision is based on the individual's decision, cf. [13]. This attests to the influence of culture on the adoptive process. In his writing on cultural relativism, Rogers hinted at some cultural categories that are normally considered part of culture: cultural relativism is 'the viewpoint that each culture should be judged in light of its own specific circumstances and needs. [...] Each culture works out its own set of norms, values, beliefs, and attitudes that function most effectively for its people' ([1], p. 441). Here, norms, values, beliefs, and attitudes are considered cultural variables that not only can be part of a community culture but also can be part of a national culture.

Historically, this aspect of DOI has been part of the anthropological research tradition in innovation diffusion research. Anthropological research has shown the influence of cultural values on the adoption of an innovation—or the failure of adoption ([1], p. 49). However, anthropology no longer played a big part in diffusion research by the end of the twentieth century ([1], p. 50).

As mentioned, with respect to (IV) nature of the social system, Rogers mentions norms and interconnectedness. In sociology, a norm is defined as 'a standard or rule, regulating behaviour in a social setting [...] dependent upon shared expectation and obligations' ([14], p. 421). Rogers writes that norms 'are the established behaviour patterns for members of a social system. [...] The norms of a system tell individuals what behaviour they are expected to perform' ([1], p. 26). Interconnectedness is the degree to which the units in a social system are linked by interpersonal networks ([1], p. 412). Both norms and interconnectedness can be viewed as aspects of an individual's social system and as aspects of a wider culture. While there can be said to be overlaps between 'social system' and 'culture', in this chapter, there is the understanding that the social system is the individual's social system and culture represents a wide group of people.

Culture is notoriously difficult to define [15]. It has been pointed out that culture represents a series of mutually incompatible concepts ([16], p. 50). With this in mind, it is imperative that the researcher is very clear about his or her understanding of culture, either by following an academic convention or by taking a substantiated standpoint (the latter is the case in this chapter).

In 2006, in a meta-analysis, Baldwin et al. [17] counted the existence of over 300 definitions of culture. The definitions range from the most restrictive (high art) to the most expansive (the totality of humanity's material and nonmaterial products) ([18], p. 11). The Baldwin et al. meta-analysis showed that researchers of culture typically have one of seven different approaches: they view culture as structure, function, process, refinement, products, group, or power [19]. This study has a structural approach to defining culture.

Kluckhohn [20] has written that 'It is important not to confuse culture with society'. This chapter also distinguishes between society and culture, that is, the understanding is that society is a wider concept than culture, a distinction that leaves political practice, economic practice, and administrative practice out of this study's understanding of culture. Consequently, this study positions itself broadly in the middle of the restrictive-expansive definitions continuum mentioned above.

According to a literature review ([21], p. 10), there appears to be two different perspectives on culture: (1) culture is pure ideology encompassing values and behaviour [22] and (2) culture is both ideology and material elements (artefacts) [23, 24]. The Baldwin et al. literature review also identified that culture includes material culture, especially with the product approach. In some definitions of culture, language is also part of culture (for instance, ([25], p. 15), ([26], p. 25), and ([27], p. 5). Samovar et al. ([28], p. 14) have pointed out that language 'transmits values, beliefs, perceptions, norms'.

The understanding of this chapter is that culture is about values, behaviour, material culture, and language. With respect to values, a distinction must be made between personal values and cultural values. Commonly, personal values are defined as the central beliefs and purpose of an individual ([14], p. 664), that is, they are at the core of what it means to be human. Some of these values appear to permeate a culture, and these aggregated values are then called cultural values. The distinction made between personal values and cultural values also applies to personal behaviour and cultural behaviour, that is, cultural behaviour is the aggregated behaviour of a population. Therefore, this chapter defines culture as being about cultural values, cultural behaviour, material culture, and language in a country.

The cultural variables identified by Rogers as part of a community culture can be categorised as follows: values, beliefs, attitudes are about cultural values; norms are about cultural behaviour. Compared with the definitions above, these four variables can only be said to represent a partial view of culture. This conclusion can make it relevant to include other aspects of culture into this investigation of culture as a determinant of innovation diffusion.

3. Literature review

Many factors seem to influence the diffusion of innovations [29]. In a review of studies, Erumban and de Jong summarised that 'the socio-cultural ambiance, perceived values, institutions and political atmosphere might influence the perception of the individuals within a society in a certain way, and these factors may consequently impact the adoption decisions' [10]. In a study of factors affecting diffusion of technology, the conclusion was that the factors that may influence the diffusion process are 'virtually limitless' [30]. Much research has focused on the economic factors, for instance, the level of income and a country's openness to trade [31] and education [32]. However, the influence of the cultural setting of a society 'has hardly received any attention in the literature' [10].

Several studies have confirmed that national culture has considerable influence on consumer behaviour [33, 34]. Studies have also told us that there are considerable variation in the rate of adoption of the same product in different countries [2]. Kumar [35] categorised studies of cultural differences in innovation into six categories, one of which was adoption of/propensity to adopt innovations. Studies that belong in this category have studied the effect of culture on various types of innovations: electronics [2], information and communications technology (ITC) [10], and innovations in general [36]. These three studies all have in common the use of the Hofstede cultural dimensions and the focus on physical goods, not services.

Dwyer et al. studied the diffusion of seven electronic innovations across 13 European countries. 'The findings suggest that national culture explains a relatively sizable amount of variation in cross-cultural diffusion rates' [2]. The research team found support linking four cultural dimensions (individualism, masculinity, power distance, and long-term orientation) to cross-national product diffusion. 'More specifically the cultural dimensions of masculinity and power distance were positively associated with the mean diffusion rate at the national level, whereas the dimensions of individualism and long-term orientation were negatively related to the mean diffusion rate' [2].

Yaveroglu and Donthu [36] studied the influence of culture on innovation and imitation (that is, will a culture, so to speak, have an innovator adopter category profile or a laggard profile) in a cross-cultural diffusion process comparison. The results indicate that cultures that are high on individualism (as is Denmark; cf. [37]), low on uncertainty avoidance (as is Denmark), and low on power distance (as is Denmark) are more likely to have a faster rate of adoption than cultures that are the opposite.

Erumban and de Jong studied the effect of culture on the adoption of ICT in more than 40 countries. They concluded that 'most of the Hofstede dimensions are important in ICT adoption. [...] In particular, the power distance and the uncertainty avoidance dimensions seem to be the most important ones' [10]. They found that countries with a low power distance score and with a low uncertainty avoidance score have a higher rate of ICT adoption (thus confirming the Yaveroglu and Donthu findings).

Many of the studies on the influence of culture on the diffusion process have used the Hofstede dimensions, and the Hofstede cultural dimensions can give answers to questions about the relationship between culture and adoption rates. However, the Hofstede cultural dimensions have also been criticised (for instance, [38, 39]), which, however, is not in itself a reason to dismiss them. It appears that the Hofstede dimensions have been used because it appears there have not been other options, or other options have been inadequate [10]. Therefore, we have to be aware that with the Hofstede dimensions, however meaningful they may be with respect to their original purpose, they are not necessarily the most meaningful to use in explaining the influence of culture on the innovation diffusion process. In theory, they could be some of the least meaningful aspects of culture to explain the influence of culture on diffusion processes. But we do not know as long as other avenues have not been explored. As Dwyer et al. [2] write 'Other approaches to culture and its measurement should be undertaken [...]'. This study will follow this recommendation and apply a different approach than dimensions of culture to explain the rate of adoption of DTT (not withstanding this position, data from research on dimensions of culture will be included in the empirical part of this study).

Three papers on cross-cultural diffusion of innovations reviewed in this section are not directly related to the present study, in that, this study is about the adoption of a service innovation, not a product innovation. This must be considered because, as Dwyer et al. write 'National culture's influence on the diffusion of service innovations should also be investigated' [2]. Since Dwyer et al. made their recommendation, not much research on this subject appear to have been carried out. Some research on the influence of culture on diffusion has been carried out with respect to specific service categories, notably internet banking services. (Banking, like DTT, is an infrastructure service (cf. [40]). It also appears that the Hofstede dimensions are used in these studies (for instance, [41]). What can be learned from these studies is that in a specific service category, there may be specific factors that can influence the rate of adoption. For instance, internet access and internet behaviour are factors that can influence the adoption of internet banking in a country [41]. Likewise, one can theorise that access and behaviour, as aspects of a culture, are factors that can influence the adoption of other service innovations.

4. Research question

This study has its focus on culture, which can be both a determinant and a barrier to the rate of adoption of innovations. In this study, the focus is on the determining factor because the case is about a 100% rate of adoption. Therefore, it does not seem obvious to focus on cultural barriers, though there may well have been cultural barriers. This observation leads to the following research question:

RQ: How can a country's culture have a determining influence on the rate of adoption of a service innovation?

The RQ mentions two variables that are related in a causal relationship: an independent variable (culture) and a dependent variable (the rate of adoption). In this study, the independent variable is being investigated both conceptually and empirically. Conceptually, it is taken into consideration that culture is a complex, multidimensional phenomenon, as mentioned above, and it is proposed that this independent variable cannot be investigated as just one variable. First of all, culture can be both a conditioning variable and a causal variable. One can say that national culture is a conditioning variable. Within a national culture, there can be certain patterns in values, behaviour, material culture, and language related to a specific phenomenon. An example could be healthcare; there can be a certain healthcare culture, for instance, trusting in religion or in evidence-based science when one is sick. In a similar vein, one can speak of a pop culture that relates to fashion, music, and consumption behaviour [42]. One can also speak of a television culture: there can be a pattern surrounding all aspects of television, both what goes on in television and what goes on in front of the television screen. Healthcare culture, pop culture, and television culture are here, as an analytical concept, called topic cultures. However, a topic culture is not separate from the national culture; it is an integrated part of national culture. With this understanding, in this study, the relationship between national culture, topic culture, and the rate of adoption of an innovation is a follows:

National culture (conditioning variable) → topic culture (determining variable) → rate of adoption (effect variable).

This study is related to a specific service offering, and the topic culture is thus the culture that relates to this specific service offering within the wider culture. The topic culture will be specified in Sections 6 and 7.

5. A case study approach

This study uses a case study approach to answer the RQ: this method is meaningful to use when the RQ begins with 'why' or, as is the case in this study, 'how' ([43], p. 8). The case study approach can be defined as follows 'A case study is an empirical enquiry that investigates a contemporary phenomenon [...] within its real-life context, especially when the boundaries between phenomenon and context are not clearly evident' ([43], p. 18). The study method relies on multiple sources of evidence ([43], p. 18). In order to study a real-life phenomenon it must have a beginning and an end. One argument for using this method is when one attempts to explain causal links in real-life that are too complex for surveys [...], in situations where a description of the real-life context is necessary ([44], p.147).

In this study, the contemporary phenomenon is the launch of DTT; the real-life context is Danish culture. It is a study of how culture may have influenced the rate of adoption of digital terrestrial television (DTT) in Denmark. DTT was launched in Denmark in March 2006, and the adoptive processed ended in November 1, 2009. Within this period, there was a 100% rate of adoption of DTT, both among households and organisations, on a nationwide scale [3]. This is considered to represent a fast process (cf. [45]), but no comparison with other cultures that are very different from Denmark appear to have been investigated with respect to the rate of adoption of DTT. Denmark was chosen because of the documentation of the rate of adoption of DTT in Denmark. With a 100% rate of adoption, the adoptive process is complete, living up to the criteria that the real-life phenomenon must have ended. This means that, with a complete rate of adoption, a discussion of determinants is more final than in an on-going process, where some unforeseen determinant can play a role.

Even though the rate of adoption was identical for both households and organisations, the types of variables and the innovation-decision processes will not have been identical ([1], Chapter 10). But what was identical was the national culture in which the two diffusion processes took place. Vejlgaard [3] suggested that Danish culture could have played a role as a determinant in the 100% rates of adoption. In another study, it was noted that that the variable that was expected to be the strongest determinant (the perception of the innovation's attributes), appears to have played a weak role in the diffusion of DTT among households in Denmark [46], indicating that other variables may have played a stronger role. One such other variable could be culture. Thus, both of these studies seem to make it relevant to examine culture as a possible determinant in the diffusion of DTT.

The case study method does not preclude the use of quantitative data; indeed, a case study can be based only on quantitative data ([43], p. 19). This study has a quantitative part.

To sum up, this study has a descriptive element as well as a causal element and is explanatory in nature. The descriptive element is Danish culture, the independent variable. The causal element is the relationship between the independent variable and the dependent variable, the rate of adoption of DTT. The causal relationship between the independent variable and the dependent variable is the study object. The dependent variable is analysed in a service perspective, and the independent variable, Danish culture, is defined with a predetermined number of variables of interest that are studied empirically using multiple sources of evidence. The unit of analysis for the empirical part of the study is the Danish population.

6. DTT in a service perspective

A starting point in this chapter is to understand theoretically what consumers may perceive they adopt when they adopt DTT. In technical terms, DTT is the transmission of television signals as digital units (bits) through the air [47]. Before DTT, there was analogue terrestrial television (ATT), which is based on broadcasting TV signals using radio waves. In popular parlance, both ATT and DTT are TV signals. As terrestrial television was already in existence, DTT is an improved TV signal. DTT allows for more TV channels to be broadcast terrestrially, in principle making it possible to launch new TV channels to television viewers who receive television signals via antenna. However, there is no guarantee that DTT leads to more (gratis) television channels being in the air. If no new channels are launched, the main benefit to viewers is a better television screen quality (higher resolution), but this should be viewed as an improvement in the television set, which also is likely to improve the television watching experience for many consumers. But a television watching experience (experience understood as the outcome of watching television) is not something that can be adopted. However, experience services (cf. [48], p. 12), such as going to the zoo, to the cinema, or to an amusement park, are all service offerings that can be adopted (or rejected). A subscription service to a cable channel or a free 'subscription' to public service television channels can also be adopted (or rejected).

The next step is to establish if DTT has a fit with an established service sector. Service management categorises services into service sectors and service industries. Guile and Quinn [40] have made an overview of service sectors and service industries. One sector, the infrastructure services sector, consists of the communications, transportation, utilities, and banking industries. Television is part of a communications industry subcategory, the broadcasting industry, which offers different types of services: TV content production, TV channels (broadcasters, the curators of TV content), and TV signal distribution companies (with distribution taking place via different platforms: cable, satellite, terrestrial, online). From an industry point of view, terrestrial television is a distribution service.

In New Service Development (NSD), a service innovation is defined as new and/ or improved service offerings, service processes, and service business models [49]. Even though the service has existed as ATT, DTT is, per the definition of a service innovation, a service innovation, even though it is 'only' an improved service. In a NSD context, one can also speak of DTT as a service improvement and, thus, as a new service (cf. [50], p. 4).

It has been pointed out that 'Service managers have difficulty in describing their product' ([51], p. 18). If service managers have difficulty in describing a service, it is not likely to be less difficult to consumers. Consumers are not likely to have a precise grasp of the television industry and know which companies produce and curate content (sometimes it is the same company) and which companies that distribute the TV signals through various platforms and by different brand names. A point here is that consumers may not perceive, understand, or experience DTT as a (new) way of distributing TV signals. However, by applying the concept of a service package, it may become clearer how consumers may perceive DTT. As with all service offerings, DTT can be understood as a service package (cf. [52], pp. 75–79). However, a service offering can be packaged in different ways. There is no definitive way to define a service package; in service literature, two service management academics, among others, have presented service package models: Grönroos [48] and Fitzsimmons et al. [51]. Their models not have different features but also have some components in common. As the Fitzsimmons model includes the experience of the consumer, the service package model by Fitzsimmons et al. is used to analyse DTT in detail. The Fitzsimmons service package model is defined 'as a bundle of goods and services with information that is provided in some environment' ([52], p. 18).

The Fitzsimmons service package model is graphically rendered as an onion model, with the (core) service experience in the middle; the explicit service and the implicit service are placed in a circle around the (core) service experience; and any supporting facility, any facilitating goods, and information are placed around the explicit and implicit services. **Table 1** is a description of each of the seven elements of the Fitzsimmons service package model with respect to DTT.

From the service provider's point of view, the core service experience is the actual service that is provided. In **Table 1**, the core service experience is defined from a typology of core service experiences [53]. However, as has been pointed out, the consumer's view may be different from that of the service provider. From the consumer's point of view, the explicit service may be the service that the consumers perceive as being delivered. The explicit service is here defined from a typology of experiences, based on experience economy theory. In experience economy theory, there are four domains—entertainment, aesthetic, education, and escapism—based on a two-dimensional matrix: customer participation (passive or active) and environmental relationship (absorption or immersion) ([54], pp. 45–47). Watching television is considered a passive experience and an experience that is absorbed by the viewer. This type of experience is defined as entertainment. For the consumer, the explicit experience of DTT is that of being entertained by watching television.

Service Pacage models elements	DTT service package
Core service experience	Enabling (cf. [53], p. 33): being able to watch television
Explicit service *The benefits that are readily observable by the senses and that consist of the essential or intrinsic features of the service.*	Being entertained by (more) television channels. Television belongs to the entertainment domain (cf. [54], pp. 45-47).
Implicit service *Psychological or emotional benefits that the customer may sense only vaguely or the extrinsic features of the service.*	Emotional benefits may be para-social (cf. [55], p. 230) or social experiences; an extrinsic feature may be social status or being informed.
Supporting facility *The physical resources that must be in place before a service can be delivered.*	In this case, a technological production facility and a network of terrestial television masts must be in place.
Facilitating goods *The physical goods needed to access the service.*	Antenna and television set.
Information *Any type of information and knowledge that is necessary to use the service.*	Knowledge on how to receive TV-signals and how to use a television set.

Table 1.
Charactristics of the service package model exemplified with DTT.

By using service theory to define DTT, a more nuanced view of what DTT may be in the mind of the consumer appears. It is likely that consumers may view DTT as a service that enables them to be entertained. This insight is important to better understand the new service that the consumers adopt when they adopt DTT. If consumers view DTT as a new service that delivers more entertainment content, in a better television screen quality than before, this may then influence how they view the benefits of the implicit service: television has several emotional benefits, most notably television offers para-social or social experiences ([55], p. 230). Extrinsic features may be about social status or about being informed and having something to talk about. These benefits do not change with DTT, and if they were important to the consumer with ATT, they are likely to be important with DTT.

How a service is perceived by the consumer may affect the motivation to adopt this service. If DTT is perceived as a part of or being connected to an experience service, the motivation to adopt may be higher than if it is perceived as 'just' technology. There may be different perceptions of DTT among the consumers, but in one way or another, many consumers are likely to relate DTT to being entertained by watching television.

When perceived as part of or connected to an experience service, DTT becomes more a part of cultural life than if it is perceived purely as technology: watching television is a cultural habit that takes place in a culture. In this way, one can also say that television plays a certain role in a culture.

The proposition in this study is that a country's culture may affect the rate of adoption of a service innovation, no matter how the service innovation is perceived; but if the service innovation is perceived as part of or connected to a service offering that is broadly founded in the culture and if the service innovation is perceived to be more about entertainment than technology, there may be a stronger motivation to adopt the service innovation.

7. Selection of cultural variables

As per the recommendation mentioned in the literature review, this study takes a novel approach to examining the relationship between national culture and the diffusion of an innovation; one that is not based on data from the Hofstede studies.

This is not without peril: as has been pointed out, it is well-known in academia that culture is difficult to define [15]. However, this study concurs with Pizam that culture is an umbrella term for a very large number of categories of phenomena ([56], p. 393). These phenomena have also been termed cultural variables or cultural factors by Smith ([57], p. 72). Some of these variables were identified in Section 2: norms, values, beliefs, and attitudes. However, it was also pointed out in this section that these variables only represent a partial view of culture. Therefore, other variables must be investigated before selecting the variables that are to be included in this study to represent the definition of culture utilised in this study.

A question is, how many variables can be said to represent key aspects of a culture? There is no objective answer to this question. However, the following guidelines will be used to determine the number of variables: the selected variables must represent all four main cultural categories. The actual number of variables is determined by the number of cultural variables in a literature review of cultural variables in the social science approach to study culture, carried out independently of the present study by Reisinger and Turner [21], as this literature review appears to overlap with cultural variables mentioned in Section 2. Approximately 50% of the cultural variables included in the literature review will be considered an adequate number of variables to represent key aspects of a culture. Also, the selection of variables must be relevant and meaningful with respect to the study object.

According to the literature review, the following variables are representative of different academic definitions of culture: culture is a way of life of a particular group of people [58, 59]. Culture indicates a pattern of social interaction [60]. Culture is a collection of beliefs, habits, and traditions [61]. Culture is the socially acquired ways of feeling and thinking [62–64]. Culture is about mental rules [60]. Culture is a system of knowledge [65]. Culture is a system of symbols and meanings [64, 66, 67]. Culture is the sum of people's perceptions of themselves and of the world [68]. Culture refers to morality, tradition, and customs [68]. Culture is about social interactions, rules about behaviour, perceptions, thoughts, language, and nonverbal communication [69]. Culture is about values, norms, customs, and traditions [22]. Culture is a combination of ideological and material elements, for instance, food, clothes, and tools [23, 24].

The literature review mentions 21 variables at least once. Therefore, the number of variables that must be selected to represent key aspects of a culture is, as a minimum, 10 variables.

The analysis of the service innovation revealed that the explicit service is about being entertained by (more) television channels. This will be input for the cultural variables to be selected for the empirical part of this study. The variables selected do not have to be the same for the national culture and for the topic culture. In this case, with respect to topic culture, the selected cultural categories should give insight into television culture, both with respect to what goes on in the television and what goes on in front of the television screen. Concretely, this can be about the types of television sets and the various television services, policies affecting television viewing, the aggregated television habits of the Danish population, and the television content.

Television belongs in several cultural categories in this literature review overview. First of all, television is about a television set, which is a designed object, belonging in the cultural artefacts category. The behavioural categories are important because watching television is mainly about behaviour. Four behavioural categories have been selected: way of life, rules, habits, and traditions. Way of life, habits, and traditions are selected because they broadly represent key behavioural aspects of culture. Rules are selected to understand if Danish culture is about following the rules of society (or not), which can have an influence on the propensity to adopt a new service that has been decided by Parliament. Television content typically represents certain values in a society, and it may be relevant to examine if what goes on in society is reflected in what goes on in television. Therefore, beliefs, thoughts, and feelings are included in the analysis. Because the language of the television programmes can have an influence on many aspects of culture, the variable language is included in the analysis. (In Denmark foreign television programmes are subtitled, not dubbed.)

All in all, nine variables have been selected. However, these variables need to be defined in order to be applied in this study: beliefs are about being sure that someone or something exists or that something is true (religion). We can have thoughts about many parts of society; here thoughts are considered ideology, which is about what is considered good and bad (politics). Ways of feelings can be about how you feel about your fellow man (social feelings) and your country (national feelings). Way of life is synonymous with lifestyle, and in this context lifestyle is based on a consumer behaviour approach to lifestyle, defining lifestyle as activities and interests (activities, for instance, work, holiday, sport; interests, for instance, home, family, recreation) ([70], pp. 171–172). Rules are prescribed behaviour. Habits are acquired behaviour that has become nearly or completely involuntary. Traditions are inherited behaviour (more formal behaviour than informal customs). Tools are physical objects broadly defined. Language is here a language group. With these interpretations and definitions, 10 cultural variables have been selected. In the present analysis, the proposition is that the following selection of variables represents the conditional variable and/or the determining variable:

- *Cultural values*: beliefs, ideology, social feelings, and national feelings.

- *Cultural behaviour*: lifestyle, rules, habits, and traditions.

- *Cultural artefacts*: tools.

- *Language*: language group.

With this selection of variables, all four main cultural categories will be included in the investigation, representing different aspects of national culture and of topic culture. The values of the cultural values will vary: in some categories, the value may be qualitative; in others, quantitative. When quantitative, it has to be established what the typical value of each of the selected cultural categories is with respect to the Danish population. This also means that what is studied are collective variables. Here, the term 'typical value' is a numerical value, defined as '50% or more' of the population. The population of Denmark must in one way or another give answers to one or more questions that can be categorised in relative terms. The answers that represent 50% or more of the population will be considered as typical in this study.

If the analysis using these 10 variables does not result in any meaningful findings, the analysis can be carried out with other variables from the literature review overview. If the inclusion of these variables also does not give any meaningful results, this study's approach to study culture as a determinant in innovation diffusion has failed and is not useful for giving insight into the national culture and television culture of a country.

8. Methodology

The overall methodological approach of this chapter is the case study approach, a qualitative approach. As outlined above, this study focuses on examining the independent variable, Danish culture, understood simultaneously as a conditional variable and as a determining variable. Studies of culture, historically within anthropology, have been qualitative, with a focus on participation-observation and other qualitative methods. However, often qualitative social research does not have a broad and truly representative sample population, which is also one of the criticisms of some qualitative research. This study seeks to counter this critique by gathering quantitative data (statistics) and using numerical values of the selected cultural variables.

As the DTT switch-over process took place in 2006–2009, quantitative data on Danish culture circa 2006–circa 2009 must be gathered. Investigating any possible cultural determinants of the rate of adoption of DTT in Denmark was not considered at that time. Almost a decade later, there are, therefore, three options: (A) gather primary data on the 2006–2009 culture of Denmark retroactively, (B) utilise existing data on the 2006–2009 culture of Denmark, and (C) do not investigate. The latter option would terminate any further inquiry under the premise presented above and leave us without insight that is lacking in research. Gathering primary data on the culture of 2006–2009 retroactively would not be credible. Therefore, option B is chosen. In this study, the existing data will be document data.

Documents have a variety of forms, for instance, books, charts, newspapers, institutional reports, survey data, and various public records [71]. The analytic procedure then 'entails finding, selecting, appraising (making sense of), and synthesising data contained in [the] documents' [71]. One can add to that 'evaluation'. As has been pointed out as a general observation, 'Adopting explicit evaluation criteria increases the transparency of [the] research and provides [...] the means to highlight the strengths and limitations of [the] research' ([72], p. 303). It has been pointed out by Bowen [71] that it is not ideal to rely on document data, but as Merriam [73] has pointed out, in some cases it is the only realistic option. Bowen writes that documents 'may be the most effective means of gathering data when events can no longer be observed or when informants have forgotten the details' [71].

Document data are typically qualitative; however, the extracted data in this study are mainly quantitative (statistical data). When utilising statistical data from multiple document sources, one has to be critical of the sources in order to secure validity and reliability ([44], p. 73). In qualitative studies, a distinction between qualitative validity (general validity) and statistical validity (number of respondents) is often made. In this study, in a first methodological stage of the research process, the statistical validity and reliability refer to the statistics used as input. The statistical input data must in one way or the other have response categories where one of the response categories represents at least 50% of respondents of a representative sample. With this methodology, the unit of analysis is de facto the Danish population.

It is here assumed that culture changes slowly, that is, culture does not change from year to year, but rather from decade to decade or over longer periods of time. Therefore, data that represent aspects of culture at various times in the period 1995–2015 are considered having validity for the time period when DTT was introduced in Denmark. The population sample must be representative of the Danish population. The method used for data gathering must be documented and live up to well-established data gathering criteria, for instance, with respect to response rates, as this affects reliability. With high statistical validity and high reliability, higher credibility is also secured, which is a key issue in all qualitative research.

Otherwise, the criteria for selecting the sources for the empirical part are:

- The source must contain quantitative data pertaining to one or more of the selected cultural variables.

- The quantitative data must in one way or the other have response categories where one of the response categories constitutes at least 50% of respondents.

- The data must be representative of the Danish population. If not, this must be clearly stated.

- The method used for data gathering must be documented in connection with the publication of the data, unless it is an opinion poll.

- If the source is an opinion poll, it must be based on a randomly selected sample that is representative of the Danish population, carried out by a professional polling organisation, using generally accepted methods of data collection, with the data having been published by the requestor, with the name of the polling organisation clearly stated.

- Data must be valid for the period 1995–2015. If several sources are available, the one closest to 2009 will be chosen.

9. Data gathering

The following document types were gathered and included in the study, as they live up to the selection criteria:

Scholarly research articles and books: [74–78]. *Academic reports and books*: Bille et al. [79] and Bonke [80]. *Master thesis*: Stephensen et al. [81]. *Cross-cultural surveys*: two cross-cultural surveys are used as sources: (1) these data were collected over multiple years, among employees working for IBM. For Hofstede's most recent book, data was collected after 2001. However, scores in this study are from the website, not from the books (the scores on the website were recalculated by Hofstede to fit a scale of 0–100). These data represent national culture [37]; (2) these data were collected over multiple years in international corporations [38]. Seventy-five percent of responses represent management; 25% represent nonmanagement. The sample is a minimum of 100 respondents. These data are not representative of the Danish population. *Opinion polls* (carried out by a professional opinion poll institute) [82, 83]. *Laws*: (Full-time Employee Law) [84] and (Media Responsibility Law) [85].

Miscellaneous documents: broadcasting schedules of church services broadcasted on the DR channel was counted manually for the year 2009, based on the daily newspaper Politiken (unpublished data) [86], Corruption Perception Index, [88] Survey of media development in Denmark 2009–2010) [87], Survey of media development in Denmark 2012 [89], Distribution of programme types in the public broadcasting system [90], Contract between the Ministry of Cultural Affairs and the public broadcasting system 2007–2010 [91], Survey of media development in Denmark [92], Survey of media habits in Denmark [93], and the official monitoring report on daily television viewing in Denmark [94].

All in all, the data span the period circa 1997–2014 which is within the timeframe stated in the methodology section ('circa' because some data were gathered before they were published). However, data that are collected at different times and used in the same analysis should warrant some caution: in this time period, the numerical values of all variables could have changed which can affect the validity of the data. However, this would not necessarily affect the outcome of this study. The sources may have varying credibility and reliability, but overall all sources are considered having high credibility and high reliability and are thus suitable to be included in an academic study.

10. Findings

10.1 Conditional variable: national culture

Faith: 91% of Danes say they do *not* base their daily living on religion ([74], p. 345).

Ideology: 69% of Danes think 'Freedom is more important than equality' ([76], p. 155). This will typically lead to an individualist culture. That that is the case with respect to the Danes can be confirmed by two sources: (1) a 74% score on the Hofstede individualism dimension means that Denmark is an individualist culture [37]; (2) a 67% score from the 1990s on the individualist-collectivist dimension confirms the stability of this score ([38], p. 51).

Social feelings: Trust is a feeling that is widespread in Denmark—76% of Danes have social trust, that is, trust in other people. 62% have trust in institutions ([78], p. 253).

National feelings: Danes have positive feelings about their own country—93% are proud to be Danish ([75], p. 326).

Lifestyle: The official workweek in Denmark is 37 hours [84].

Rules: With respect to following rules, 62% of Danes take the so-called universalist approach on the universalist-particularist dimension ([38], meaning they would put rules first and foremost. It is also practised on a private level: when Danes observe someone they know wanting to drive while under the influence of alcohol, 81% have acted to stop it [83]. Adherence to rules is clearly practised at the government level, as Denmark is the country in the world with the least corruption (no. 1 on the Corruption Perception Index 2014) [87].

Traditions: Danish culture is normative. Normative societies prefer to maintain time-honoured traditions and norms while viewing societal change with suspicion. People in such societies exhibit great respect for traditions [37]. Eighty-three percent of the Danes support the monarchy [82].

Language groups: Danish is the primary discourse language in Denmark.

10.2 Determining variable: television culture

Faith: There are no religious television stations in Denmark [92], and there appear to be very little religious programming in Danish television channels [86]. However, the Danish Broadcasting Corporation (DR) has an obligation as part of a so-called public service contract to broadcast some religious broadcasting [91]. DR broadcasts church services on a certain number of Sundays and religious holidays.

Ideology: Danish television channels are independent of party political influence of politicians and the government [85]. All political views are represented fairly evenly in the public service television channels [81].

In Denmark, there are no government-run television stations, though the Danish Parliament lays out some guidelines for broadcasting certain types of content for public service television stations, with a focus on Danish language programming and programmes relating to Danish society and culture. These guidelines do not apply to commercial television stations [91]. In other words, Danish television stations are independent of government authority and the government's politics, except with respect to the public television service obligations.

The programming of the two main Danish DR public service channels had the following programming content in 2009: presentation of programmes (4%), education (3%), sport (1%), entertainment (6%), music (6%), non-Danish movies and series (30%), Danish movies and series (4%), general topics and culture (27%), current news (15%), and general news (8%) [90].

Social feelings: In 2008, Danes had the highest trust in television news among 14 news sources [77].

National feelings: the ratio of Danish to non-Danish television programming varies from television channel to television channel. However, the DR public service channels have a specific obligation to broadcast programming in Danish and on Danish culture [91].

Lifestyle: In 2009, Danes, both men and women, had 8 hours of free time per day ([80], p. 11). Two hours a day are spent on practical work; 2.5 hours are spent on relaxation; 2 hours are spent on social activities with other people; the rest is mainly spent on eating and drinking [93]. All of these activities can take place in front of the television. The daily leisure activity that many Danes spend most time on is television ([79], p. 291).

Habits: Danes watch a lot of Danish television—94% and 93%, respectively, watch the two main Danish public service television channels on a daily or weekly basis ([79], p. 53). In 2009, there was a record in television viewing in Denmark: on average Danes watched 3 hours 9 minutes of television per day [88]. Television watching peaks at 9 PM when 55% watch television [93]. 65% of the population are interested in watching entertainment programmes; 72% are interested in watching movies ([79], p. 56).

Traditions: One strong tradition in Danish culture and Danish television culture is watching the Queen of Denmark's New Year speech, which is broadcasted live on the main public service television channels on New Year's Evening. Approx. 60% the Danish population watch the speech [94].

Tools: 99% of the Danish population own at least one television set ([79], p. 51). In 2009, more than 53% of the population had more than one television set in their primary home. In 2009, 56% of Danish households had at least one flat TV screen [89].

Language groups: 70% of programming on the two main DR public service channels were in Danish in 2009 [90].

11. Analysis

Television appears to be an embedded part of cultural life in Denmark. Therefore, it is likely that many aspects of national culture and of television culture could have played a part in determining the rate of adoption of DTT in Denmark: almost all Danes own one or more televisions sets, and they are used—the main measured leisure activity in Denmark is television watching. Most of the content of Danish television can broadly be categorised as entertainment. The Danes to a large extend watch television in popular entertainment categories, though categories that are not categorised as entertainment may also be perceived as entertainment by some viewers.

The television culture seems to be a reflection of the national culture: religion does not play a big part in Danish cultural life, neither in the real life nor on television. If religion played a big part of what was broadcasted, that would be at odds with the general Danish culture. (In other cultures, religion may play a big part both in the general culture and in the television content.) When there is a balance between the general culture and the television content, this is likely to have a positive influence on the motivation to adopt DTT.

In Denmark, television is independent of government interference. Politicians do not have a say over content, exempt within some broad guidelines, which matches the Danish individualist-independent culture. With respect to political content, all points of views are represented on television, which indicates that most viewers will feel that their views are represented on television.

To a large extent, television programming is based on commercial considerations and/or broad public service guidelines. Danish television stations are independent of government authority and the government's politics, which is likely to make the Danes take a positive attitude towards television. This is likely to have had a determining influence on the rate of adoption of DTT.

Trust in public institutions is high in Denmark. While trust in media is generally low, trust in television news is high. Trust in public institutions can have a positive determining influence on the rate of adoption of DTT, and the fact that television is second highest in trust can have had a determining influence on the rate of adoption of DTT.

The Queen's New Year televised speech plays a big part in the public discourse in Denmark, in the media, and in private conversations, around New Year's evening [95]. Thus, the Queen's speech can be perceived as playing a part of the implicit service of DTT: being able to take part in the private and public discourse surrounding this broadcast event. This is just one example of a television tradition, but if the Danes perceive this as an implicit service of DTT, this is likely to have had a determining influence on their motivation to adopt DTT. The existence of a tradition of watching television on certain occasions can have a determining influence on the adoption of DTT.

Finally, one can make a somewhat banal observation: Danish is spoken in a big part of the content in the public service channels. The fact that in almost all areas there is coherence between the national culture and the television culture, also with respect to language, is likely to have had a positive determining influence on the rate of adoption of DTT.

If the Danes viewed DTT as an enabling service, making them able to watch television, this is likely to have had a determining influence on their willingness to adopt DTT. Likewise, if the Danes viewed the explicit service of DTT as being entertained, this is also likely to have had a determining influence on their willingness to adopt DTT.

12. Discussion

It seems that television plays a role in Denmark that would motivate many television viewers to want to be able to watch television also when DTT would be the only option for watching television. In conclusion, one can say that both the national culture and the television culture of Denmark were conducive to adopting DTT in Denmark.

The proposition in this study has been that a country's culture may influence the rate of adoption of a service innovation, no matter how the service innovation is perceived, but if the service innovation is perceived as part of or connected to a service offering that is broadly founded in the culture, there may be a stronger causal relationship between the culture and the rate of adoption of the service innovation. This study has made the argument that adopters may have associated DTT with entertainment, which plays a big part in Danish television culture. This study did not investigate if DTT was perceived as part of an entertainment domain by Danish adopters; therefore, this cannot be confirmed empirically. But with respect to convergence, this study hints that the convergence of content and technology that takes place at the industry level may not be particularly confusing to consumers because they may not make the distinction between service/content and technology in the first place.

Ten cultural variables from a list of 21 cultural variables that have some overlap with cultural variables identified in previous research represented the framework for the study. The selected variables represented all four categories of the definition of culture. Therefore, the answer to the RQ is that cultural values, cultural behaviour, material culture, and language in unison may have a determining influence on the rate of adoption of a service innovation. It would be overstating the findings to suggest that specific variables have specific influence, but it is certainly relevant for further research to establish a ranking of the influence that different cultural variables may have on the rate of adoption of service innovations. The framework appears to be meaningful in studies of cultural determinants of the rate of adoption of innovations, not just of service innovations. However, in studies of other topics, it may be relevant to include other cultural variables than the ones used in this chapter.

Rogers categorised the determinants in five categories, based on research up to 2003. Based on the findings of this study and the other studies mentioned in the literature review, it appears meaningful to add a sixth category, (VI) culture. With this addition, it is likely that we can get more precise insight into what determines the rate of adoption of an innovation. This may then have consequences with respect to the relative influence of the causal variables. Rogers pointed out that 'Little research has been carried out to determine the relative contribution of each of the five types of variables [...] ([1], p. 223). This study cannot offer precise insight into the relative contribution of each of the now six categories of variables that determine the rate of adoption. However, it is likely that the determining variable that is the strongest predictor will vary with type of innovation, the innovation's target group and the context of the innovation, including the cultural context. This study has not generated findings on the relative influence of culture on the rate of adoption of DTT, but it is likely to have played a role. The fact that both households and organisations within the same culture had a 100% rate of adoption can hint of a not minor determining influence of culture. The relative influence of culture certainly seems relevant to include in future research.

As has been pointed out, it seems plausible that intangible service offerings may be more influenced by national culture than are physical goods and products. This study cannot establish this, as a comparison has not taken place. But a point is that a positive correlation (i.e. culture affects the adoptive process of service innovations) may be true for all service innovations, but the actual influence can be either positive (speeding up) or negative (slowing down) the rate of adoption: in the case of DTT, the television culture in Denmark is likely to have had a speeding up influence on the rate of adoption of DTT. But in a country, where television does not play a strong role in the culture, this may have a slowing influence on the rate of adoption of DTT. Thus, it may not be the national culture as such that has the biggest cultural influence but rather the topic culture. Therefore, the concept of topic cultures is also relevant to include in future DOI research.

As was pointed out by Fitzsimmons et al. [51], companies may have difficulty in putting into words what their services are. This may or may not be the same for consumers. But the difference is that consumers do not have to put this into words; they just have to react to an innovation that may influence a big part of their everyday life. Especially when the innovation in question is a service improvement of an existing service, the consumers have insight into at least some features of the innovation, and they know if they need the service or not. When the innovation is a service that is important in their daily life (their media habits) and the cultural life of the country, it is not likely to be a question of *if* they want to adopt but a question of *when*. By making the distinction between national culture and topic culture, we are likely to get better not just at understanding a causal relation but also of being able to predict the outcome of an adoptive process. The latter is a key element in diffusion of innovation research, which is not only descriptive but also predictive: we want knowledge on how to predict the rate of adoption of innovations. Thus, the findings of this study are also relevant to the forecasting research field.

Author details

Henrik Vejlgaard
Copenhagen Business Academy, Copenhagen, Denmark

Address all correspondence to: hev@cphbusiness.dk

IntechOpen

References

[1] Rogers EM. Diffusion of Innovations. New York: Free Press; 2003

[2] Dwyer S, Mesak H, Hsu M. An exploratory examination of the influence of national culture on cross-national product diffusion. Journal of International Marketing. 2005;**13**(2):1-27

[3] Vejlgaard H. Fast organizations: A comparative study of the rate of adoption in households and organizations. International Journal of Technology Diffusion. 2015;**6**(3):21-31

[4] Greenhalgh T, Robert G, Macfarlane F, Bate P, Kyryiakidou O. Diffusion of innovations in service organizations: Systematic review and recommendations. The Milbank Quarterly. 2004;**82**(4):581-629

[5] Baldwin JR, Faulkner SL, Hecht ML. A moving target: The illusive definition of culture. In: Baldwin JR, Faulkner SL, Hecht ML, Lindsley SL, editors. Redefining Culture. London: Lawrence Erlbaum; 2006. pp. 3-26

[6] Kluckhohn C. Values and value-orientations in the theory of action: An exploration in definition and classification. In: Parsons T, Shils E, editors. Toward a General Theory of Action. Cambridge: Harvard University Press; 1951. pp. 388-433

[7] Smith PB. Book review. International Journal of Cross Cultural Management. 2007;**8**(1):110

[8] Minkov M, Hofstede G. Is national culture a meaningful concept? Cultural values delineate homogeneous national clusters of in-country regions. Cross-Cultural Research. 2012;**46**(2):133-159

[9] Minkov M, Hofstede G. Clustering of 316 European regions on measures of values: Do Europe's countries have national cultures? Cross-Cultural Research. 2014;**48**(2):144-176

[10] Erumban AA, de Jong SB. Cross-country differences in ICT adoption: A consequence of Culture? Journal of World Business. 2006;**41**:302-314

[11] Hofstede G. Culture's Consequences: Comparing Values, Behaviors, Institutions, and Organizations Across Nations. Thousand Oaks, CA: Sage Publications; 2001

[12] Rogers EM, Steinfatt T. Intercultural Communication. Prospects Heights. Ill: Waveland Press; 1999

[13] Im I, Hong S, Kang MS. An international comparison of technology adoption: Testing the UTAUT model. Information Management. 2011;**48**:1-8

[14] Jary D, Jary J. Sociology. London, England: Collins; 2000

[15] Condon J, LaBrack B. Culture, the definition of. In: The SAGE Encyclopedia of Intercultural Competence. Thousand Oaks, CA: SAGE Publications; 2015

[16] Winthrop RH. Dictionary of Concepts in Cultural Anthropology. New York: Greenwood; 1991

[17] Baldwin JR, Faulkner SL, Hecht ML, Lindsley SL, editors. Redefining Culture. London: Lawrence Erlbaum; 2006

[18] Griswold W. Cultures and Societies in a Changing World. Thousand Oaks: Pine Forge; 1994

[19] Faulkner SL, Baldwin JR, Lindsley SL, Hecht ML. Layers of meaning: An analysis of definitions of culture. In: Baldwin JR, Faulkner SL, Hecht ML, Lindsley SL, editors. Redefining Culture. London: Lawrence Erlbaum; 2006. pp. 27-52

[20] Kluckhohn C. Mirror for Man. New York: McGraw-Hill; 1949

[21] Reisinger Y, Turner LW. Cross-Cultural Behavior in Tourism: Concepts and Analysis. Burlington: Butterworth-Heinemann; 2003

[22] Rokeach M. The Nature of Human Values. New York, NY: Free Press; 1973

[23] Assael H. Consumer Behavior and Marketing Action. Boston, MA: PWS-KENT Publishers; 1992

[24] Mowen J. Consumer Behavior. New York, NY: Macmillan; 1993

[25] Neuliep JW. Intercultural Communication: A Contextual Approach. Boston: Houghton Mifflin; 2003

[26] Lustig MW, Koester J. Intercultural Competence. New York: Longman; 2006

[27] Brislin RW. Understanding Culture's Influence on Behavior. Ft. Worth: Harcourt; 2001

[28] Samovar LA, Porter RE, Jain NC. Understanding Intercultural Communication. Belmont, CA: Wadsworth Publishing Company; 1981

[29] Wejnert B. Integrating models of diffusion of innovations: A conceptual framework. Annual Review of Sociology. 2002;**28**(1):297-326

[30] Rosenberg N. Factors affecting the diffusion of technology. Explorations in Economic History. 1972;**10**(1):3

[31] Caselli F, Coleman WJ. Cross-country technology diffusion: The case of computers. American Economic Review. 2001;**91**(2):328-335

[32] Lee JW, Barro RJ. Schooling quality in a cross-section of countries. Economica. 2001;**68**(272):465-488

[33] Ganesh J, Kumar V, Subramaniam V. Learning effect in multinational diffusion of consumer durables: An exploratory investigation. Journal of the Academy of Marketing Science. 1997;**25**(3):214-228

[34] Roth MS. The effects of culture and socioeconomics on the performance of global brand image strategies. Journal of Marketing Research. 1995;**32**(2):163-175

[35] Kumar V. Understanding cultural differences in innovation: A conceptual framework and future research directions. Journal of International Marketing. 2014;**22**(3):1-29

[36] Yaveroglu IS, Donthu N. Cultural influences on the diffusion of new products. Journal of International Consumer Marketing. 2008;**14**(4):49-63

[37] Hofstede. 2016. http://geert-hofstede.com/denmark.html [Accessed: December 15, 2016]

[38] Trompenaars F, Hampden-Turner C. Riding the Waves of Culture: Understanding Cultural Diversity in Business. London: Nicholas Bealey Publishing; 1997

[39] Yoo B, Donthu N, Lenarttowiccz T. Measuring Hofstede's five dimensions of cultural values at the individual level: Development and validation of CVSCALE. Journal of International Consumer Marketing. 2011;**23**(3-4):193-210

[40] Guile BR, Quinn JB. Technology of Services: Politics for Growth, Trade, and Employment. Washington, DC: National Academy Press; 1988

[41] Takieddine S, Sun J. Internet banking diffusion: A country-level analysis. Electronic Commerce Research and Applications. 2015;**14**(5):361-371

[42] Storey J. Cultural Theory and Popular Culture: An Introduction. London: Routhledge; 2018

[43] Yin RK. Case Study Research: Design and Methods. London: Sage Publications; 1989

[44] Johns N, Lee-Ross D. Research Methods in Service Industry Management. London: Cassell; 1998

[45] Vejlgaard H. Late adopters can be fast: The case of digital television. Communications. 2016;**41**(1):87-98

[46] Vejlgaard H. Rate of adoption determinants of innovations: A case study of digital terrestrial television. International Journal of Digital Television. 2018;**9**(1):7-26

[47] Benoit H. Digital Television: Satellite, Cable, Terrestrial, IPTV, Mobile TV in the DVB Framework. Oxford: Focal Press; 2008

[48] Grönroos C. Service Management and Marketing. New York: Wiley; 2007

[49] Johnson S, Menor L, Roth A, Chase R. A critical evaluation of the new service development process: Integrating service innovation and service design. In: Fitzsimmons J, Fitzsimmons M, editors. New Service Development: Creating Memorable Experiences. Thousand Oaks, CA: Sage Publications; 2000

[50] Fitzsimmons J, Fitzsimmons M, editors. New Service Development: Creating Memorable Experiences. Thousand Oaks, CA: Sage Publications; 2000

[51] Fitzsimmons JA, Fitzsimmons MJ, Bordoloi SK. Service Management: Operation, Strategy. In: Information Technology. New York: McGraw-Hill International; 2014

[52] Normann R. Service Management: Strategy and Leadership in Service Management. New York: Wiley; 2007

[53] Bryson JR, Daniels PW, Warf B. Service Worlds: People, Organizations, Technologies. New York: Routhledge; 2004

[54] Pine BJ II, Gilmore JH. The Experience Economy. Boston: Harvard Business Review Press; 2011

[55] McQuail D. Mass Communication Theory: An Introduction. London: Sage Publications; 1989

[56] Pizam A. Cross-cultural tourist behavior. In: Pizam A, Mansfeld Y, editors. Consumer Behavior in Travel and Tourism. New York: The Haworth Hospitality Press; 1999

[57] Smith SE. Effects of cultural factors on mass communication systems. In: Speech Communication Association, editor. International and Intercultural Communication Annual V. Falls Church, VA: Speech Communication Association; 1979. pp. 71-85

[58] Harris M. The Rise of Cultural Theory. New York, NY: Crowell; 1968

[59] Harris P, Moran R. Managing Cultural Differences. Houston, Texas: Gulf Publishing Company; 1996

[60] Harris M. Cultural Anthropology. New York, NY: Harper and Row; 1983

[61] Mead M. Cultural Patterns and Technical Change. Paris, France: UNESCO; 1951

[62] Harris M. Culture, People, Nature: An Introduction to General Anthropology. New York: Harper and Row Publishers; 1988

[63] Nisbett RA. The Social Bond. New York, NY: A. A. Knopf; 1970

[64] Radcliffe-Brown AR. A Natural Science of Society. Glencoe, Ill: Free Press; 1957

[65] Keesing RM. Theories of culture. Annual Review of Anthropology. 1974;**3**:73-97

[66] Kim YY, Gudykunst WB. Theories in Intercultural Communication. International and Intercultural Communication Annual, 12. Newbury Park, CA: Sage Publications; 1988

[67] Schneider D. Notes toward a theory of culture. In: Basso K, Selby H, editors. Meaning in Anthropology. Albuquerque, NM: University of New Mexico Press; 1976

[68] Urriola O. Culture in the context of development. World Marxist Review. 1989;**32**:66-69

[69] Argyle M. The Psychology of Interpersonal Behaviour. Harmondsworth, England: Penguin Books; 1967/1978

[70] Kotler P. Marketing Management: Analysis, Planning, Implementation, and Control. New York: Prentice Hall; 1991

[71] Bowen GA. Document analysis as qualitative research method. Qualititative Research Journal. 2009;**9**(2):27-40

[72] Eriksson P, Kovalainen A. Qualitative Methods in Business Research: A Practical Guide to Social Research. London: Sage; 2015

[73] Merriam SB. Case Study Research in Education: A Qualitative Approach. San Francisco, CA: Jossey-Bass; 1988

[74] Borre O. Politiske værdier. In: Gundelach P, editor. Danskernes særpræg. Hans Reitzels Forlag: Copenhagen; 2004

[75] Hobolt SB. Lokal orientering. In: Gundelach, Danskernes særpræg. Copenhagen, Denmark: Hans Reitzels Forlag; 2004

[76] Juul S. Solidaritet. In: Gundelach P, editor. Danskernes særpræg. Hans Reitzels Forlag: Copenhagen; 2004

[77] Schrøder K. Danskerns brug af nyhedsmedier: Et nyt landkort. Journalistica. 2010;**1**:8-37

[78] Sørensen JFL, Svensen GLH, Jensen PS. Der er så dejligt derude på landet? Social kapital på landet og i byen 1990-2008. In: Gundelach P, editor. Danskernes særpræg. Hans Reitzels Forlag: Copenhagen; 2004

[79] Bille T, Fridberg T, Storgaard S, Wulff E. Danskernes kultur- og fritidsvaner 2004 – med udviklingslinjer tilbage til 1964. Copenhagen: AKF Forlaget; 2005

[80] Bonke J. Er fritiden forsvundet? 45 års forskning i danskernes fritid. Copenhagen: Rockwool Fondens Forskningsenhed; 2014

[81] Stephensen M, Christoffersen A, Olesen AS. Frem til fortiden, partipressens genkomst? [master's thesis]. Institute of Political Science: University of Copenhagen; 2006

[82] Berlingske. Danskerne knuselsker monarkiet. Copenhagen: Berlingske; 2013

[83] Rådet for Sikker Trafik. Press Release: Spritkørsel er upopulært. Copenhagen: 2013. www.sikkertrafik.dk

[84] Funktionærloven. https://www. retsinformation.dk/Forms/R0710. aspx?id=179871 [Accessed: March 20, 2018]

[85] Medieansvarsloven. https://www. retsinformation.dk/Forms/R0710. aspx?id=143047 [Accessed: March 20, 2018]

[86] Broadcasting schedules. 2009. Unpublished raw data

[87] Transparency.org http://www. transparency.org/news/feature/ corruption_perceptions_index_2016 [Accessed: March 20, 2018]

[88] Medieforskning
DR. Medieudviklingen 2009-2000-2015
2010. Copenhagen: DR; 2011

[89] Medieforskning
DR. Medieudviklingen 2012.
Copenhagen: DR; 2013

[90] DRs public serviceredegørelse 2010.
Copenhagen: DR; 2010. https://www.
dr.dk/NR/rdonlyres/8DB7AE5A-20EA-
420B-91F6-CCE1EF79CBD8/6121053/
Public_serviceredeg%C3%B8relse_2010.
pdf [Accessed April 2, 2018]

[91] Kulturministeriet. Public Service
Kontrakt 2007-2010. Copenhagen:
Kulturministeriet; 2006

[92] Kulturstyrelsen (2015). Mediernes
udvikling i Danmark. Copenhagen:
Kulturstyrelsen. [Accessed: April 3,
2018]

[93] TNS Gallup. Multi Medie
Mennesket 2009. Copenhagen: TNS
Gallup; 2009. http://www2.tns-gallup.
dk/media/132714/m3_praesentation_
mkl_111109.pdf [Accessed: Arpil 3,
2018]

[94] TNS Gallup. Seertal uge 53 (28
December 2009-3 January 2010). 2009.
http://tvm.gallup.dk/tvm/pm/2009/
pm0953.pdf [Accessed: April 3, 2018]

[95] Author's observation, 2000-2015

Chapter 5

Television as a Surveillance Tool

Ananda Mitra

Abstract

"Television" has now become a screen for the projection of different kinds of data that operate in a full-duplex fashion where the screen and its accessories not only provide data to the viewer of the screen but also become a tool for the collection of data about the viewer. For instance, the moment a smartphone is connected to the screen to watch an episode on YouTube, data about the viewer and viewing habits of the viewer are available to a corporation. This process opens up significant opportunities for data collection that borders on surveillance, and different institutions are able to collect customized information about the individual. This chapter will explore the mechanisms of this process of surveillance and the benefits and burdens offered by this emergent technology.

Keywords: television, big data, surveillance

1. Introduction

Television (TV) that became a ubiquitous part of households in many parts of the world since World War II has been witnessing a significant transformation since the turn of the twenty-first century. Starting in the early 2000s, TV has morphed in many ways such as in the size and quality of picture it delivers, the kind of content it can offer, and the multiple ways in which it can be used.[1] Yet, in spite of these changes, it has been the case that TV has remained a site to consume narratives.[2] Within the narrative paradigm, narratives, as pointed out in the work of Fisher [7–9] and later in the research on narrative bits [10–13], are also windows to the lives of individuals and groups. While TV has brought narratives home, the knowledge of the narratives of people can allow one to better understand the person and predict and control what the person may do. This perspective on narrative suggests that every person has a "life story," and access to that story offers an insight into the person's life. The challenge has been accessing the story in detail. Creating a detailed narrative requires constantly watching the person and tracking the person's beliefs, interests, and behavior. The matter of watching and constructing the narrative was eased when the advent of the digital allowed the analog, flesh-and-blood person, to construct a digital representation of the self in the digital space. This was akin to creating the life story online, which could be the repository of the narrative of the person. Indeed, this is the realm of big data [14, 15]. In this essay I argue that TV, originally the conduit for offering passive narratives to the audience, is

---

[1] There are numerous books on the history of television and its development that are used for introductory courses in mass communication that enumerate this aspect of the development of television.

[2] The focus on narratives in television has been examined by many scholars who make the argument that television is eventually about story telling [1–6].

transforming into a tool that can watch over the audience and construct a dynamic narrative of the audience, thus operating as tool for surveillance.

2. The passive narrative

Since the early days of TV in the developed countries of the West, technology and medium had been considered to be a passive device that was the conduit that brought information to the people who would watch the screen seeking anything from entertainment to education. TV has sometimes been called the "idiot tube," for the mesmerizing effect it would have on the watcher who could be distracted to catatonic inactivity just watching TV in a "mindless" way without having to bring any intellectual energy to the process of watching TV. This phenomenon was examined copiously by scholars from many disciplines, and numerous theories were proposed and debated that examined the "effects" of watching TV as would be found in many introductory books on mass communication.

One important assumption that underpinned the emergent theories claimed that the audience of TV was a relatively passive and often disengaged person (see, e.g., early research by Klapper [16]). This assumption was especially true for the programs of research that mimicked the natural scientific methods of research devising experiments and interventions with samples selected from the population to understand the effects of TV in numerical terms, as in the case of development of theories such as cultivation theory often considered to be a central tenet of understanding the effect of TV. Other researchers who subscribed to a more cultural anthropological, critical, and cultural interpretations of the role of TV in everyday life sought answers in the ways the audience would talk about television or through observational studies where the audience would be observed to see how they interacted with the narratives and discourses on television, as in the case of scholar originating in the Birmingham Centre for Cultural Studies and offering the vast array of literature on the role of TV within popular culture starting with the work of scholars such as [17]. In both approaches, however, there was the shared presumption of relative anonymity of the audience where any individual member of the audience was a part of a larger similar kind of people where the specific individual was unknown to those who created and circulated the content of TV. This presumption worked well for the industry because the content producers were only concerned with creating content that could be of appeal to a certain type of audience and not to any individual person since there was no definitive way of knowing who the person was.

This lack of information about a specific member of the audience was largely a factor of the way in which TV technology worked from its inception to the time when the Internet became a part of everyday life for large groups of people. Traditional TV technology was designed to deliver a robust image and sound to the audience without the audience having to come to the place where the content was available as in the case of movies. Like its predecessor—radio, TV brought the message to the home of the audience. The content distributors had little knowledge of who was watching the content, why they were watching the content, or if the audience was liking the content. For the content distributors such as the NBCs and BBCs, once the content left the antennas, there was no way to "control" the content and trace where it went or what happened to the content. At the reception end of the process, TV technology was a "passive" tool that merely displayed the content on the screen. Once the TV was turned off, the screen was just a part of the furniture in the room. This status quo changed with the increasing adoption of the Internet in the public sphere.

For a length of time, the television screen achieved a sense of status quo until around the 1980s when one of the key quests was to push the size of the screen so that a cinema-like experience could be reproduced in the privacy of homes for those who could afford the huge back-projection units that often had pictures of poor quality. This trend to improve the picture quality continued for decades.

However, a change started to happen in homes of the developed nations in the latter part of the 1990s, and by the 2000s, there was an increasing interest in a different screen that had made its way into the households of the developed nations—the computer monitor—very similar in technology to TV but often only available for displaying text that would appear on the screen in monochrome. However, the magic of the computer screen was in the fact that there was an additional device, the computer, which was connected to the screen that allowed the user of the computer screen to interact with the screen unlike the user of TV screen who merely viewed the screen. This change was especially important, because the interaction produced an active audience who could personalize the experience of using the screen. Even if the use was restricted to typing words on the screen, it was a different form of interaction with a device that looked similar to TV screen that the user was already accustomed to.

This interactivity with the computer screen progressed in several different directions in the early part of the twenty-first century. With increasing home-based access to the digital network of global computers—the Internet—the interactive computer screen became a conduit to a larger virtual space with increasing libraries of data that the user could access. This data, often residing on computers all over the world, could be accessed by any of the computers using the computer screen. Even though the computer and the TV screen were beginning to look similar, their functions were constantly diverging with the TV screen becoming a site for narratives that the user could not control. The narratives of the TV screen were simply sent out to the user with the expectation that the user would subscribe to the narratives when the screen brought them home; indeed, the users were expected to manage the everyday life practices of their lives to suit the demands of the TV screen if the users wanted to access the narratives on the TV screen. Therefore, people would plan their evenings around the shows they would watch on TV [2]. On the other hand, the narratives were within the control of the user on the computer screen. Here, the user could build an unending narrative by accessing multiple data that were connected by hypertext to each other allowing the user to constantly explore, discover, and construct the personalized narrative that the user sought and not what the TV institutions handed out.

The seduction of interactivity, coupled with the primacy of the computer screen over the TV screen, led to the demand for a single-screen solution where the screens could be merged into one where the single screen would serve primarily as a conduit for interactions that would allow the user to construct their personalized narrative that would appear on this single converged screen. It is this demand for convergence that allowed the ubiquitous merged screen to become a site for collecting data about individuals.[3]

3. The interactive narrative

The notion of convergence is particularly important in the context of the emergent screen in the private spaces occupied by individual members of the audience. The duality between computer screen as a site with the potential of creating an

---

[3] The notion of the screen has a multifaceted implication as discussed elsewhere [18].

interactive narrative, such as writing a book, and the TV screen as a site of consuming narrative was increasingly being erased as a single converged screen was replacing the two where the single screen would converge the different functions into one site. The notion of technological convergence precisely states that new tools often diminish the need for multiple tools with multiple functions into a single tool that offers the convenience of doing many functions with one gadget.[4]

The new digital TV with access to the Internet built into TV was becoming commonplace by the early 2000s and became nearly ubiquitous within 5–6 years, especially in the USA where all TV broadcast changed to digital broadcast on June 12, 2009, and nearly 97.5% of the American homes were ready for the mandatory transition.[5] Within the next several years, the transition to digital TV, the nearly ubiquitous availability of broadband connection to the Internet, and the emergence of content conglomerators, producers, and distributors that were distinct from the traditional media content providers offered new narratives to the audience. Thus, the dominance of corporations such as Netflix, Amazon, Hulu, YouTube, and others less prominent institutions allowed for narratives to be converged on the now-interactive screen in the private home. The narratives could now smoothly travel from the screen of the smartphone to the screen of a tablet computer, to the screen in the living room, to the projection system in the den, to the screen on the back of the driver seat in a car, and to the seat-back screen on a transcontinental flight. The same narrative was available everywhere.

This narrative was also partially composed by the viewer and could be completely distinct from the narratives composed and consumed by other viewers. This narrative had a distinct characteristic of interactivity that was missing in the traditional TV narrative.

The notion of interactivity, as proposed here, is derived from the way TV narratives have traditionally worked. In discussing the "flow" of television, it has been suggested that the programming decisions by the traditional TV content providers, from the CNNs to the BBCs, were thoughtfully made to construct an intertextual narrative that would span an evening of watching TV where the passive viewer has no little choice but to follow the narrative pattern constructed by a network. The availability of hundreds of TV channels, and the simple remote control, allowed viewers to "interact" with the narrative, create a partially customized narrative by switching between channels, and construct a narrative that gratified the audience, albeit within the limits of what was available on the channels. The availability of recording technology from the traditional VCR to the DVRs allowed users to shift the viewing time to one under the control of the viewer and watch only the narratives that the individual viewer was interested in.

The popularity of the Internet, accompanied with the availability of content producers and distributors mentioned earlier, however, altered the way in which the viewer could interactively construct narratives. First, it became far simpler to shift the viewing time, an advantage that was already available through the more elaborate home-based recording technologies. The viewer now had the ability to seek and find narratives at any time the viewer wanted to consume narratives. The boundaries of space and time were disrupted because the ubiquitous connectivity to the Internet through multitude of digital devices, some of which were portable, allowed the viewer to call upon programs and narratives anywhere and anytime the individual wanted. Second, the narratives could be obtained from a multitude of sources where the viewer was no longer restricted to the traditional providers of narratives such as the TV channels. The increasing digitization of video (and

---

[4] https://www.sciencedirect.com/science/article/abs/pii/S0308596198000032

[5] https://www.nielsen.com/us/en/insights/news/2009/the-switch-from-analog-to-digital-tv.html

audio) allowed for narratives to be obtained from sources that would never be considered providers of narratives, including noninstitutional sources that could not have afforded to be in the public sphere before the availability of the Internet. In particular, YouTube is the example of the worldwide video-sharing platform. A viewer could now call upon narratives that were from individual composers of narratives who would never be found in the traditional media spaces. All that was needed was the ability to do the appropriate queries to yield the kinds of narratives that the viewer was interested in. Third, the viewer could interact with multiple sources of narratives and create a customized "playlist" that specifically would be designed to meet the interest of the viewer and could be distinct from other viewers. Even though the viewer was still restricted to the narratives that were connected to the network, the choice was sufficiently large that a viewer could construct a very specific playlist to satisfy the "taste" of the viewer. Finally, all of the narratives, and the queries that create the conglomeration of narratives, could now be done through the interface of the TV screen which transformed from the passive screen to a site of interaction between narratives and the viewer.

The common theme for this transformation is the interactive power attributed to the viewer with the privilege of being able to search for the narratives based on the interests of the viewer. It is precisely this interactivity, now happening through the press of buttons on a TV remote control, that transforms the relationship between the viewer and the TV screen which now serves as the gateway for the vast digital space where content is located. It is precisely the nature of the gateway that makes the TV screen the window into the world of the individual viewer who, while watching the narratives, is also being "watched" by the TV screen.

4. Watched by TV

In February 2018, an analysis by the reputed magazine *Consumer Reports* announced that their testing revealed that the increasingly ubiquitous "smart TV" was capable of "watching" the viewer and keeping a detailed record of the viewer's TV watching patterns and related behavior.[6] As more of smart devices find a place in the average home, there are other gadgets that can work in tandem with smart TVs to perform the task of "watching." Consider, for instance, the Alexa devise that responds to voice commands to perform simple tasks, including connecting with a smart TV to control the smart TV.[7] All such devices and functions rely on the fact that these devices always "surveil" its environment—watching with built-in cameras, listening with built-in microphones, and capturing data with built-in sensors. Real people occupy the space that is under the surveillance of these devices.

It is useful to briefly consider the way in which the process of surveillance has been examined over a period of time. The practice of surveillance has been around since the times that people wanted to "watch over" others. The need to watch has most importantly been related to the notion of security where the watcher has been concerned about the fact that the watched poses a threat to the interests of the watcher. Those interests could be intertwined with the interests of the watched as well; thus, the process of watching becomes particularly important to maintain a sense of order within a specific societal system. Indeed, this perspective was aptly summarized by Mike Rogers, the chairman of the intelligence committee in the American House of Representatives, following the embarrassing report in 2013 that

---

[6] https://www.consumerreports.org/televisions/samsung-roku-smart-tvs-vulnerable-to-hacking-consumer-reports-finds/

[7] https://www.zdnet.com/article/how-cia-mi5-hacked-your-smart-tv-to-spy-on-you/

the National Security Agency (NSA) was surveilling the phone conversations of European leaders such as Angela Merkel. Mr. Rogers was quoted to have said, "It's a good thing. it keeps the French safe. It keeps the US safe. It keeps our European allies safe." [19]

The intimate connection between the maintenance of order and discipline becomes the central thesis of the academic examination of the process of surveillance when scholars such as Foucault [20] begin to connect surveillance to power and discipline. Among the different ideas of surveillance that emerged as important was the notion of the Panopticon which claims that the powerful is constantly watching everything all the time [21]. The Panopticon society was built around a strict definition of discipline, and in the late 1800s and early 1900s, the metaphor was principally used to describe the ways in which totalitarian nations and despots would want to constantly watch everything to maintain power and discipline (see, e.g., [21–27]).

In some cases, however, there is the emergent interest in examining how the watchers could also include corporations and institutions that had a motive unrelated to discipline and power but more interested in understanding the "market" that the institution would be interested in serving (see, e.g., [28]). This is especially true for the type of interactive technologies described in this essay. The advent of the technologies described earlier in this essay is, however, concerned with the corporate watching rather than the discipline- and power-based Panopticon world that earlier scholars were concerned with. TV in the house is now constantly watching and monitoring the individuals that use TV not to stop sedition or to exercise power over the watched but to better understand the "taste" of the watched to ensure that the watcher can best deliver content to the watched that the watched is most likely to consume. In a transactional system where commodities would be sold for profit the process of TV watching, the audience is to better commodify the audience who can then be sold to appropriate institutions as a part of a potential market. The point of interest in this transaction is not the seditious behavior of the individual, as in the case of cameras watching for shoplifters in large shopping areas, but more in constructing the life story of the individual to analyze and predict what the individual may like to consume. The process of watching is thus tied to creating the life story of the audience that TV can obtain by "watching" the data that the individual generates. The data was being generated for a long period of time through a variety of digital tools that a person could be using, but TV converged all the functions of data collection into one console which increasingly becomes ubiquitous in the life of most individuals in the developed and developing worlds. The Panopticon TV in the living room is thus watching a set of different things that early surveillance studies have pointed toward, albeit no longer in the context of discipline and power.

The new Panopticon created by TV at home is however less about discipline and power and much more about the way in which the "customer" who is being watched can be analyzed as a commodity who can be sold to those that are interested in selling to the watched. Simultaneously, the Panopticon condition becomes far more benign and perhaps even comforting to the watched by creating a cocoon of comfort within which the watched can dwell, where the cocoon is created by the TV itself. This process is possible because the customer voluntarily interacts with the TV by offering information to the TV and the vast array of interests that the TV represents. There are broadly two kinds of information that the watched offers to the watcher through the modern television—attitudes and behavior.

The information about the attitudes, interests, beliefs, and tastes is offered by the specific discourse the watched offers to the different providers of information that bring content to the TV. Consider, for instance, the simple act of accessing a digital video service such as YouTube that can be accessed on a smartphone and then projected on the TV. In some cases, the TV itself would offer the option of connecting directly to a service such as YouTube. Indeed, it is estimated that nearly 80% of TVs in American homes would be connected to the Internet by 2019 and any TV that is connected to the Internet can potentially be accessing YouTube without the need for any other ancillary device.[8] This connection makes TV the conduit for the vast amount of data available on YouTube as well as many other segments of the digital space that contain searchable data. One of the key aspects of this connection is the ability of the person being watched search for specific kind of content that can be accessed by TV and displayed on the screen. The person inscribes attitudes and preferences in the language of the search. Companies like Google have been using similar information for a long time and are thus able to offer personalized advertising when a person is working on a computer. There are ways in which such personalization of marketing messages can be turned off through the adjustment of specific settings on an application provided by a corporation. The matter becomes a little different on TV where the very purpose of the tool, the TV, is to watch narratives, and in the environment of services such as YouTube, the viewer must reveal interest information to customize what the person is watching or interested in watching. The process of using TV to access narrative content is intimately connected with the process of revealing to TV the watcher's interests, attitudes, and beliefs.

This information is also connected with the disclosure of behavior patterns. Given that much of the consumption of the content is happening through the content providers such as YouTube, Hulu, Netflix, and other Internet-based content delivery systems, there is a constant record of what was watched, when it was watched, how it was paid for, and in some cases greater granular information related to the particular watcher in a multi-people home. For instance, Netflix offers the opportunity to set up multiple subaccounts under one primary account for each member of the household, and the data that is built up actually shows which particular person was actually using specific content. In homes that have multiple TVs, it is also possible to surveil which particular TV was being used to watch what content offering a detailed understanding of the specific members who are being watched by the corporations through the conduit of TV.

The attitude and behavior data that such surveillance offers eventually become a narrative about the people who are being watched over. It is this narrative that becomes especially important in the new Panopticon system produced by the modern TV.

5. The story of the watched

As suggested in the opening of this essay, analog person has increasingly been supplemented by the digital self where the latter can be constructed as a story about a person using the data that is produced by the analog being. The surveillance that the TV does within the privacy of the home is geared toward the construction of that narrative. A specific and unique narrative is produced by the Panopticon TV which examines the different aspects of the life of a person, and this life story of the person becomes a part of the analog person itself. This narrative can be quite detailed with some specific characteristic because of the number of different aspects of a person's life that is being watched by the TV as indicated earlier.

---

[8] https://www.telecompetitor.com/report-percentage-of-smart-tvs-actually-connected-to-the-internet-on-the-rise/

First, the narrative is cumulative. TV is constantly watching and updating the narrative. Every time an individual interacts with the TV, a new segment is being added to the narrative of the life of the person. This process is similar to the way in which other data about an individual is constantly updated, as in the case of the combination of location based on Global Positioning System (GPS) and applications that offer mapping information such as Google Maps. These applications retain the records of the movement of a cell phone through space and are thus constantly updated offering a "time line" of spaces that a person might have inhabited.[9] The TV surveillance operates in a similar way because the attitude and behavior data being collected by the TV is also constantly updated and the ongoing narrative of the life of a person is stored for future reference. This certainly has its advantages, where a viewer can, for instance, resume watching a show from where it was left off, offering the Panopticon TV an opportunity to see how the "rhythm" of a person's life unfolds on a moment-to-moment basis. Similarly, because TV knows the story of a person's life, it knows, through its applications, what the person may like to watch next with very well throughout suggestions being offered by the TV with respect to what entertainment the watched individual may be encouraged to watch. The TV watcher's life story is now known to the TV, and TV can gently help to shape that story to reinforce the elements of the story that have been prominent over time. Thus, a person who watched a few episodes of a science fiction would be encouraged to watch other shows belonging to the same genre.

The longitude of the narrative is also connected with the way in which an attempt is constantly made by the different tools of surveillance, including TV, to triangulate the data to create a narrative about the individual which would encompass *all* the data about the person. Current laws may make it a little difficult to correlate all the data sets, as in the case of privacy laws in the USA where the medical data of an individual is held sacrosanct and unavailable and generally unconnected with other narrative elements of a person's life. However, there is sufficient data about a person that can be available to the Panopticon TV which would allow the TV to surveil the individual in a more precise manner and further help design the ongoing narrative of the person being watched. For instance, gadgets like the Alexa, which respond to voice commands, can be connected to the TV to control the TV with spoken words, as explained in a guide, "Once you link Quick Remote with your Roku device and Alexa, you can use voice commands to tell Quick Remote to navigate the Roku menu system and select any app to start playing.[10]" There are two important aspects that need to be noted in these instructions: first, it shows how to connect three different applications (Quick Remote, Roku, and Alexa) to each other to have the convenience of sending voice commands to the Panopticon TV. All these three systems are sharing the data with each other and thus creating a robust narrative about the person who is being watched.

The second important aspect pointed out in these and numerous other such instructions is that the user, or the watched, is offering the data to construct the narrative. There are no hidden cameras or stealthy sensors that are surreptitiously watching the person. On the other hand, the person chose to find the convenience of talking to the TV and thus voluntarily obtained the devices and the applications which help to create the dynamic narrative that eventually makes the life of the person more comfortable. Indeed, this comfort is best maintained if the person's life story is fully known to the Panopticon TV and its army of other devices that is constantly updating the narrative of the person and creating the zone of comfort for the person that eventually becomes comfort for the analog self where the digital

---

[9] https://www.wired.com/story/google-location-tracking-turn-off/

[10] https://www.lifewire.com/use-alexa-with-your-tv-4161152

narrative helps to predict what the analog self needs. Consider, for instance, the notion of Internet of Things (IoT) that hopes to convert information from and about every device that surrounds an individual to a centralized interconnected database about the person making the life story as complete as possible. When such projects come to fruition, the surveillance, aided by the voluntary data offered by the individual, would transcend the TV. In that future, all devices, including TV, would be geared to collating the most complete life story of the watched.

6. Where does this leave us? The watched

There are a few things worthy of note with respect to the way in which TV has transformed into a tool for watching the watcher. First, this process has not been forced upon a population who had no option but to be watched. A small amount of knowledge about the ways in which the tools are watching us can allow us to shut off the surveillance. None of these tools, including TV, makes the data collection process a "required" activity to use the tool in its basic and rudimentary way. One can certainly watch television shows broadcast "over the air" without connecting the TV to the Internet. In a similar way, it is possible to use the Alexa speaker as only a portable speaker connected to a smartphone that has music stored in it. Indeed, even a smartphone can be used to make phone calls only without connecting it to the digital realm.

However, as these examples show, when a user chooses to not connect the TV to the Internet, or Alexa to its manufacturer, and the smartphone to a data plan, the user is sacrificing the ability to use the tools to their full potential. Additionally, the user is sacrificing access to the numerous programming options offered through these tools. There is, therefore, a constant tension between the inclination to maintain a sense of privacy while watching TV and retaining the convenience of the TV making suggestions about what would be interesting to watch. If TV is allowed to surveil, and it is connected with the other tools that surround the TV, then it will eventually be able to create an increasingly complete life story of the person who uses TV. This complete life story could become the way in which TV constructs a mediated reality for the person who is being watched. As discussed earlier, this reality can become progressively myopic and an echo chamber within which the person would reside while the Panopticon TV creates the comfortable media space for the person.

This future is increasingly realistic since the function of TV as the bearer of the programs offered "over the air" or even through the cable system that became commonplace in the 1980s is quickly shifting. In many parts of the world, there is the increasing tendency to "cut the cord" and get rid of the cable delivery of programming. Cable companies are increasingly facing a threat where the centrality of program delivery by cable is being replaced by program delivery via the Internet. Numerous companies such as Amazon, Roku, and Apple are offering accessories that could be connected to the TV, and program would be delivered through the connection of the accessory to the Internet. Thus, a Roku "stick" can connect to the Internet, and the programs would be offered by Roku in collaboration with other content aggregators such as Sling, YouTube, and Hulu, to name a few. In some cases, a complete ecosystem is produced by a company like Amazon that would offer the accessory for TV, a household voice activated information retrieval system such as Alexa, and content through the vast store of content that Amazon owns. As the user is migrating to these options, the user is also required to share information through the conduit of the TV with all these different corporations that continue to watch the watcher. It is indeed a world of constant surveillance, whenever the TV is switched on.

Author details

Ananda Mitra
Department of Communication, Wake Forest University, Winston-Salem, USA

*Address all correspondence to: ananda@wfu.edu

IntechOpen

References

[1] Feuer J. Narrative form in American Network Television. In: MacCabe C, editor. High Theory/Low Culture. New York: St. Martin's; 1986. pp. 101-104

[2] Fiske J. Television Culture. New York: Methuen; 1987

[3] Fiske J, Hartley J. Reading Television. London: Methuen; 1978

[4] Hall S. Encoding/decoding. In: Hall S, editor. Culture, Media, Language. London: Hutchinson; 1980. pp. 128-139

[5] Kozloff S. Narrative theory and television. In: Allen R, editor. Channels of Discourse Reassembled. Chapel Hill: University of North Carolina Press; 1992

[6] Metz C. Film Language: A Semiotics of the Cinema. New York: Oxford UP; 1974

[7] Fisher WR. Narration as human communication paradigm: The case of public moral argument. Communication Monographs. 1984;**51**:1-22

[8] Fisher WR. The narrative paradigm: An elaboration. Communication Monographs. 1985;**52**:347-367

[9] Fisher WR. The narrative paradigm: In the beginning. Journal of Communication. 1985;**35**:74-89

[10] Mitra A. Creating a presence on social networks via narbs. Global Media Journal. 2010;**9**(16). http://www.globalmediajournal.com/open-access/creating-a-presence-on-social-networks-via-narbs.pdf

[11] Mitra A. Narbs as a measure and indicator of identity narratives. In: Dudley et al., editors. Investigating Cyber Law and Cyber Ethics: Issues, Impacts and Practices. Hershey, PA: IGI Global; 2012

[12] Mitra A. Mapping narbs. In: Wise G, editor. New Visualities, New Technologies: The New Ecstasy of Communication. New York, NY: Ashgate Publishing Ltd; 2013

[13] Mitra A. Digital DNA: Managing Identity in Social Networking Sites. New Delhi, India: Rupa Publications; 2014

[14] Bughin J, Chui M, Manyika J. Clouds, big data, and smart assets: Ten tech-enabled business trends to watch. Financial Times. 2010

[15] Frank AD. IBM CEO Rometty says Big Data are the next great natural resource. The Daily Beast. 2013

[16] Klapper JT. The Effects of Mass Communications. Oxford, England: Free Press of Glencoe; 1960

[17] Hoggart R. The Uses of Literacy. New York: Routledge. 1998

[18] Mitra A. India on the Western Screen. SAGE Publications. 2016

[19] Sherwell P, Barnett L. Barack Obama 'approved tapping Angela Merkel's phone 3 years ago'. The Telegraph. 2013

[20] Foucault M. Discipline and Punish: The Birth of the Prison. New York: Advantage Books; 1979

[21] Bentham J. The Panoptic Writings. London: Verzo; 1995

[22] Bentham J. Principles of penal law. In: Bowring J, editor. The Works of Jeremy Bentham. Vol. I. New York: Russell and Russell; 1962. pp. 365-580

[23] Bentham J. The Rationale of Evidence. In: Bowring J, editor. The Works of Jeremy Bentham. Vol. I. New York: Russell and Russell; 1962. pp. 201-585

[24] Gandy O. The Panoptic Sort: A Political Economy of Personal Information. Boulder, CO: Westview Press; 1993

[25] Lyon D. The Electronic Eye: The Rise of Surveillance Society. Minneapolis, MN: University of Minnesota Press; 1994

[26] Lyon D. Surveillance as Social Sorting. New York: Routledge; 2003

[27] Lyon D. Theorizing Surveillance: The Panopticon and Beyond. Devon: Willan; 2006

[28] Manokha I. Surveillance, panopticism, and self-discipline in the digital age. Surveillance and Society. 2018;**16**(2):219-237

www.ingramcontent.com/pod-product-compliance
Lightning Source LLC
Chambersburg PA
CBHW081239190326
41458CB00016B/5840